SUPERVISORS SAFETY MANUAL

SUPERVISORS SAFETY MANUAL

Better production without injury and waste from accidents

NATIONAL SAFETY COUNCIL, CHICAGO, ILLINOIS 60611

FIRST EDITION © 1956

SECOND EDITION © 1961

THIRD EDITION © 1967

FOURTH EDITION

COPYRIGHT © 1973 NATIONAL SAFETY COUNCIL

ALL RIGHTS RESERVED

*No portion of this book may be reproduced by any
process without written permission of the National Safety Council,
425 North Michigan Avenue, Chicago, Ill. 60611*

LIBRARY OF CONGRESS CATALOG CARD NUMBER: 73–77916

INTERNATIONAL STANDARD BOOK NUMBER: O–87912–061–4

PRINTED IN THE UNITED STATES OF AMERICA

60M87308   Stock No. 151.01

PREFACE
TO THE 4th EDITION

The key man in an occupational hazard control program is the first-line supervisor. Not only is he the direct link between top management and the work force, but on his shoulders rests the responsibility for quality job training, development of good safety attitudes, and detection of unsafe conditions and practices.

Not only must he know the techniques of human relations, but he must know the fundamentals of accident prevention. He himself must be trained if he is to review hazard control information with his workers, check personal protective equipment and safety devices and investigate accidents occurring in his area.

This Supervisors Safety Manual is designed to provide this safety training. The Manual reflects the experience of many engineers of the National Safety Council's Industrial Department who have worked with supervisors and safetymen in setting up effective safety programs and in using this book as a text in NSC courses.

The Manual is now in its fourth edition. Many changes and additions have been made since the first edition was published in 1956. The National Safety Council appreciates the help of the many supervisors, safety directors, and other professionals who gave information and advice in preparing these editions. New information and many new illustrations have been added to this latest edition. Chapters 3, 6, and 11 were revised extensively.

To increase the effectiveness of the Manual, much detailed material has been included. References to Occupational Safety and Health regulations refer to those published in Title 29, Chapter XVII of the *U.S. Code of Federal Regulations,* revised to October 18, 1972 (published in Vol. 37, No. 202, of the *Federal Register*). However, it

v

is not the purpose of this Manual to serve as a complete handbook; it attempts, rather, to set forth the important points to be considered by a safety-minded supervisor. In fact, some of the recommendations given here exceed present OSHA requirements.

The information and recommendations contained in this Manual have been compiled from sources believed to be reliable and to represent the best current opinion on the subject. No warranty, guarantee, or representation is made by the National Safety Council as to the absolute correctness or sufficiency of any representation contained in this and other publications, and the National Safety Council assumes no responsibility in connection therewith. Nor can it be assumed that all acceptable safety measures are contained in this (and other publications), or that other or additional measures may not be required under particular or exceptional conditions or circumstances.

Staff credits for the fourth edition of this Manual are as follows: Overall direction was provided by Leonard C. Smith, manager of the Training Department. Individual chapters were revised under the supervision of these NSC staff men: Chapter 1, John Mark; Chapter 2, John DuFon; Chapter 3, Leonard Smith; Chapter 4, Alfred Finch; Chapter 5, Leonard Smith; Chapter 6, Ed L. Alpaugh, P.E., industrial hygienist, International Harvester Co., Chicago, Ill.; Chapter 7, Ralph Coe; Chapter 8, Charles Price; Chapter 9, Frank Parker; Chapters 10 and 11, John Epperson; and Chapter 12, Joe VanSickle. Editing and production were headed by Frank McElroy, director of Technical Publications Division, and a registered professional engineer.

CONTENTS

1

Safety and the Supervisor
Cost of Accidents: Accidents and efficiency, Accidents and morale, Accidents and public relations. Growth of the Safety Movement. Safety—The Supervisor's Job: The supervisor's safety program.

1

2

Know Your Accident Problems
See and Know Job Hazards: Correct first things first, Review department experience, Systematic inspection, Legal requirements for safety, Information sources. Elements of an Accident: Unsafe conditions and unsafe acts, Accident investigation. Measuring Safety Performance: Accident costs, Recordable injuries and illnesses, Incidence rates.

11

3

Human Relations
Background of Motivation: Principles of motivation, Knowing that work has value, Knowing that work is appreciated, Having a sense of security, Belonging to the work group. The Supervisor as Leader. Preparing the Worker for Changes. Workers with Special Problems: Temporary problems, Talking reduces strong emotions, the "accident-prone" individual. The Alcohol and Drug Problem: Interaction of alcohol and drugs, Alcohol and drugs on the job, Countermeasures. Summary.

38

vii

Contents

4

Maintaining Interest in Safety
How To Promote Safe Workmanship: Employee committees, Safety posters, Special-purpose reminders, Off-beat ideas, Recognition organizations, Safety contests, Suggestion systems, First aid courses, Safety meetings. Off-the-Job Accident Problems: Cost of off-the-job accidents, Humanitarian reasons for OTJ safety activities, Measuring accident experience, Integrating OTJ with on-the-job safety programs, The supervisor's role in OTJ safety.

57

5

Instructing for Safety
Giving Instruction: Importance of job instruction, Starting the new man. On-the-Job Training: Over-the-shoulder coaching. Job Safety Analysis: Select the job, Break the job down, Identify hazards and potential accidents, Develop solutions, Benefits of JSA. Job Instruction Training: Step 1—preparation, Step 2—presentation, Step 3—application, Step 4—testing, Followup. Other Methods of Instruction: The lesson plan, Programmed instruction, Independent study, Closed-circuit TV, Summary.

82

6

Industrial Hygiene
Chemical: Injestion, Absorption, Inhalation. Physical Classification of Airborne Materials: Dusts, Fumes, Smoke, Aerosols, Mists, Gases, Vapors, Hazards. Solvents: Selection and handling, Degree of severity of solvent hazards. Particulates: Dispersion, Nonmetallic dusts, Fumes. Physical Agents: Noise, Ionizing radiation, Nonionizing radiation. Visible Radiation: Lasers. Temperature Extremes: Sweating, Radiation, Conduction, Effects of high temperature, Heat controls. Atmospheric Pressures. Ergonomics: Mechanical vibration, Repeated motion. Biological stresses: Tuberculosis, Fungus infections, Byssinosis, Anthrax, "Q fever," Brucellosis, Erysipelas, Upper respiratory tract infections. Threshold Limit Values: Establishment of values. Dermatitis: Occupational acne. Standard Operating Procedures: Entering tanks, Maintaining pumps and valves, Industrial hygiene controls.

106

7

Personal Protective Equipment

Controlling Hazards. Protecting the Eyes: Equipment types, Face protection, Overcoming employee complaints. Protecting the Feet and Legs: Leg protection. Protecting the Head: Fabrication, Auxiliary features, Overcoming objections. Ear Protection: Amount of protection, Insert-type protectors, Muff types. Protecting Fingers, Palms, and Hands. Protecting the Torso. Respiratory Protective Equipment: Selecting the respirator, Types of equipment, Cleaning the respirator, Overcoming employee complaints. Safety Belts and Harnesses: Fabrication, Lifelines, Inspection and care of belts and lines. Protection Against Ionizing Radiation: Monitoring radiation. Safe Work Clothing: Materials for protective clothing, Clothing for women, Special problems today for men. "Selling" Personal Protective Equipment: Paying for protective equipment.

151

8

Industrial Housekeeping

Results of Good Housekeeping. Supervisory Responsibility: Clean up, Directing others. Sources of Help: Safety department, Maintenance department, Planning department, Purchasing department, Methods department, Plant fire protection. Inspections: Techniques, Plant structure, Floors, Aisles, Storage facilities, Employee facilities, Machines and equipment, Yards and grounds. Color.

201

9

Materials Handling and Storage

Materials handling problems. Manual Handling Methods: Lifting and carrying, Handling specific shapes. Equipment for Handling: Hand tools, hand trucks, Powered hand trucks, Powered industrial trucks, Dock plates, Conveyors, Cranes, Railroad cars, Highway trucks, Motorized equipment. Rope, Chains, and Slings: Fiber ropes, Rope slings, Wire rope, Chains and chain slings, Storing chains. Materials Storage: Containers and other objects, Hazardous liquids, Gas cylinders, Combustible solids.

224

Contents

10

Guarding Machines and Mechanisms
Principles of Guarding: Benefits of good (safe)-guarding, Definitions. Part I—(Safe)guard Design: A (safe)guarding problem, Maximum size openings, Types of (safe)guards, Built-in (safe)guards. Part II—(Safe) guarding Mechanisms: Rotating mechanisms, Cutting or shearing mechanisms, Inrunning nip points, Screw or worm mechanisms, Forming or bending mechanisms, Primary and secondary operations, Feeding methods, Ejecting material, Foot controls. Part III—(Safe)guard Types and Maintenance: Fixed guards or enclosures, Interlocking guards or barriers, Automatic protection devices, Automation, Safe practices, Maintenance of (safe)-guards, (Safe)guarding for maintenance and repairs, Lockout procedure, Replacements, (Safe)guarding hazards before accidents.

281

11

Hand and Portable Power Tools
Control of Tool Accidents: Centralized tool control, Personal tools, Carrying tools. Maintenance and Repair: Tempering tools, Safe-ending tools, Dressing tools, Handles. Use of hand tools: Metal-cutting tools, Wood-cutting tools, Miscellaneous cutting tools, Torsion tools, Shock tools, Spark-resistant tools. Portable Power Tools: Selection, Electric tools, Air power tools, Special power tools.

329

12

Fire Protection
Basic Principles: Understanding fire chemistry, Determining fire hazards, Informing the working force. Causes of Fire: Electric equipment, Friction, Special fire-hazard materials, Welding and cutting, Open flames, Portable heaters, Hot surfaces, Smoking and matches, Spontaneous ignition, Static electricity. Fire-safe Housekeeping. Alarms, Equipment, and Evacuation: Fire alarms, What about extinguishers? Followup for fire safety, Fire brigades, Special fire protection problems.

369

Useful References **399**

Index **401**

CHAPTER 1

Safety
and the Supervisor

Every safety and health program has a single purpose—performance of activities without accident, injury, or occupational illness.

Meeting this challenge in our rapidly expanding and changing technology can eliminate not only the tragedy of human suffering and death, but also the high costs, waste, and poor quality that result from accidents. Added benefits from such efforts aid in the compliance with state and federal codes and lessening the burdensome troubles of medico-legal problems.

It disturbs us to see one of our own employees get hurt. Not only do we feel sorry for the fellow and his family, but we often get angry when it looks as though the fellow was not concerned about his own safety. We often feel bad because we did not foresee that something might happen and then take steps to prevent it.

"But," we say to ourselves, "we can't think of everything, or be everywhere all the time. If we spend every minute trying to keep people working safely we will slow down production."

We, as supervisors,* want to prevent accidents, but also must want

---

* For purposes of discussion in this book, the terms "foreman" and "supervisor" are used interchangeably to mean one who immediately directs the job efforts of a working group.

1

"We nearly had an accident!" Even if there is no personal injury, accidents *are* accidents. They are expensive and add to cost of production. Left photograph shows javelin-like projectile hurled by circular saw blade through half-inch-thick plywood. At right, poor stacking leads to spilled coils and damaged stock.

to spend the necessary time to do it.

Look at it this way. Accident prevention and efficient production go together. Production is higher and quality is better when accidents are prevented. Actually, it takes less time and costs less money to prevent accidents than it does to have them.

Safety, therefore, is not something for us to work at when we have the extra time. Supervision and safety go hand in hand. Implementing the company program—making sure his work area is safe and that his people work safely—is an integral part of the supervisor's total responsibility.

Cost of Accidents

Accidents cost money—and result in higher costs, less profits and lower quality. Money lost because of accidents is not like money spent for material or wages. There is no return on money expended on accidents.

Think about how this money is dissipated.

Everyone can see money spent for medical expenses and for workmen's compensation. And, although the amounts vary, in no case do dollars fully compensate an employee for his lost time, his lost working capacity, and his suffering.

Even if a company carries insurance to take care of losses, the company still eventually pays for them. The greater a company's injury experience, the higher are the costs. Insurance premiums must cover losses. Every U.S. state and Canadian province has a compensation law. These laws are constantly being liberalized—both minimum and maximum payments, as well as the scope of misfortunes covered, are increasing.

Medical, compensation, and insurance costs are a necessary cost of doing business. Insurance premiums, however, do not pay the entire accident bill. What should be realized is that the hidden costs of accidents—those which are not covered by insurance—can be higher than medical, compensation, and insurance costs. The rule-of-thumb figure is that hidden costs are at least equal to and in some cases more than the direct cost. (The next chapter discusses this in detail.)

Property damage is one of these hidden costs. Often an accident occurs which injures no one (or in which the injury is slight), but which does extensive property damage. For example, a crew unloading a truck neglected to secure the dock-plate that extended from the dock to the truck platform. Heavy power trucks passing over it moved the plate a fraction of an inch each time until it was supported only on the very edge of the dock. The next power truck pushed it off and crashed down between truck and docks—fortunately, the operator came out on top and was only scratched a little.

The company accident report showed only a nondisabling injury with first aid treatment and no lost time. But repairs to the power truck cost several hundred dollars. Because this bill was charged to maintenance and repair of equipment, it never appeared in the accident reports. Also, production time was lost while the damaged truck was removed. Yet it was an accident cost just as surely as compensation paid to an injured worker.

And what about the near-accident?

"We nearly had an accident!" the foreman of a woodworking

3

machine room reported to the safety director. "A helper fed a short piece of maple into a molding machine designed to take only long pieces. The short piece got free of the feed rolls, jammed into the high-speed cutter head, and broke the knives. They flew out like bullets. Luckily, no one was hit," the foreman emphasized.

Extensive damage was done to the machine, and production was delayed. But the foreman only reported: "We *nearly* had an accident."

Actually, it was an accident. Accidents do not have to result in injury. A failure can occur without causing an injury.

Accidents without personal injury occur frequently. They have the same causes as personal injury accidents, and they can be prevented in the same ways. They are expensive and add to the cost of production. Preventing these accidents is important. Accident prevention is just as much the foreman's responsibility as is production.

Accidents and efficiency

All accidents reduce efficiency and effectiveness. They are, furthermore, symptoms that something is wrong. Because accidents stem from lack of control over men, materials, processes and environment—lack of control will inevitably detract from an efficient and effective operation.

A good production man likes to operate on schedule. He wants to know that he has a man available to handle each job. He wants to know each day how close each job is to completion. Efficient production demands effective planning.

To accomplish this, the supervisor himself needs to know what he is going to do next. He needs time for planning, time for improving methods, time for figuring out other tasks.

An accident adversely affects the operation of a system. It is never "scheduled" and often happens at the most inconvenient time. It may knock out one of your best producers, or damage key equipment. At the very least, it is sure to pull the supervisor away from his regular responsibility—not only to take care of the injured man, but also to arrange for a temporary (and sometimes permanent) replacement. If equipment has been damaged, repairs or replacement must be scheduled. Other production and other departments may be affected. The product may be damaged, and have to be repaired or scrapped.

The supervisor should investigate every accident, including those not resulting in personal injury, if he is to solve his accident problems. A battery-operated recorder helps speed up the investigation.

This affects delivery dates.

When any of these results occur, the supervisor realizes the importance of an effective safety program.

It is not just the big accident that causes trouble. A series of little accidents can keep a department, or even a company, in a state of inefficiency.

The supervisor, generally, must investigate accidents and make reports. Both duties take time. Often this time does not show up on cost sheets so nobody is concerned about it—except the foreman. If, however, he spends too much time "cleaning up" accidents, production begins to suffer.

Accidents and morale

A general once defined morale as "the belief by every man that his regiment is the best in the army, that his company is the best in the regiment, and that he is the best man in the company." No soldier, however, can feel that way if his leader makes frequent mistakes. Neither can a worker have that feeling about his job if his department has a high accident rate.

A high injury rate depresses employee morale. Unsafe plant conditions contribute to accidents, and lead to lowered morale and poor job performance.

A bad accident makes everyone nervous, sometimes fearful. Repeated accidents make employees feel their company does not care about them, or feel that their supervisor is not on top of his job. They lose confidence. They lose interest in doing their job well.

An outstanding safety record, on the other hand, contributes to real pride and enthusiasm, just like an outstanding production record does. Employees develop loyalty to their company and a sense of a job security.

Good morale is worth more than money in the bank. A foreman who builds morale in his men renders important service to his firm. He is doing his job.

Accidents and public relations

Building a good reputation in the community is second only to building a good reputation among employees. (Also see page 76.)

A good safety record is noticed by the entire community. A company gets to be known as being "a good place to work." Management values this kind of reputation and thinks well of supervisors who help gain it. Bad accidents, too, are reported to the community, and damage a company's reputation.

Every employee, every activity, every facility of a company contributes to the overall feeling that persons outside a company have about that company. This is true public relations.

Prevention of accidents, therefore, must not be a sideline, followed when there is time or when it is convenient. Accident prevention is part of business—it pays off in good morale and in good public relations and image.

Growth of the Safety Movement

Before 1910, few companies had established safety programs, much less some form of compensation payments to injured workers. Companies did not do much about accident prevention because they thought there was not much they could do. They looked upon accidents as being the fault of careless workers or as occurrences that could not be avoided. The accident situation got worse. The predicament of injured workers and their families became so bad that it became a social issue.

As a result, states began to pass workmen's compensation laws—the first state to pass the law was Wisconsin in 1911. Today every state and Canadian province has such a law. Although the laws differ in details (including even the amount paid for a compensable injury), they all require an employer to pay stated amounts to workers injured while at work, regardless of who was at fault.

It is doubtful that those who drafted the laws knew what would be the full effect of their action. Although the law required payments for injured workers, the more important result was that employers began to take steps to prevent accidents.

Active safety work in most companies began shortly after the laws were passed. The first step they took was the most obvious—to safeguard machinery. From 1911 to 1920, progress in safeguarding machinery was great. The number of crippling injuries was reduced sharply, proving the success of the program. But this was only the beginning.

A cooperative Safety Congress was held in Milwaukee in 1912, at the National Convention of the Association of Iron and Steel Electrical Engineers. Industry, insurance companies, and government agencies joined in this historic meeting. As a result of this meeting, the National Safety Council was formed.

In those days, Council membership consisted of a few companies who were actively engaged in accident prevention. But membership grew rapidly, and today the Council is made up of thousands of member companies in every kind of business. Not only business firms, but associations, schools and colleges, and local, state, and federal government agencies join in this international safety effort. Most members of the National Safety Council are in the United States and Canada, but

7

many companies in foreign lands are also represented.

The Council publishes magazines and books, produces films and other training aids, provides consulting services, and conducts the National Safety Congress, attended by more than 13,000 persons every fall in Chicago. The Council staff of several hundred professional safety people and specialists coordinates the work of over a thousand committeemen from member organizations. The best ideas from all over the world are pooled to make this the greatest cooperative safety effort in the world.

In 1970, Congress passed the Williams-Steiger Occupational Safety and Health Act. This Act, known as the OSHAct, requires every employer to provide a safety work place for employees. Inspections of facilities by federal compliance officers are made to make sure that provisions of the Act are complied with. Penalties are provided for noncompliance.

The supervisor's responsibility to effectively supervise for both production and safety takes on greater importance.

Safety—The Supervisor's Job

Once safety programs got rolling, they could not be stopped. Soon safety people discovered that many accidents had nothing to do with unguarded machinery. They found that the way people worked, and the care that they and company management took, were also important factors. Tremendous campaigns were undertaken to make people "safety conscious"—safety committees of workers and management were organized, mass meetings were held, contests were conducted, and posters were displayed. All these measures were effective, and the number of injuries kept coming down.

Too often, however, accident prevention was regarded by management and workers as a sort of campaign that was added to the job, but not really part of it. Safety committees and safety departments took the responsibility for safety. About the most foremen did was participate as safety committee members, and even then, not often.

Gradually, management realized that safety was not part-time work but an important part of business that could not be separated from the other parts—that doing a job correctly was doing it safely—and it had to be done at the same time. They found that one of the

The supervisor contributes to the safety program by training employees in proper work practices, by supervising and evaluating their performance, by adhering to all safety rules and regulations, and by constantly monitoring the workplace for unsafe conditions and reporting them to management.

biggest parts of the accident prevention job belongs to the foreman.

Preventing accidents is tied up closely with many of the important duties a supervisor normally has, even when there was no formal safety program. Here is a summary of these duties:

Maintain order

Keep work on schedule

Maintain equipment

Establish work methods and procedures (often in cooperation with specialists)

Instruct workers

Keep men busy (no idle time)

Supervise work

Adjust complaints

Maintain morale

Control costs

Control quality

Assign jobs

No supervisor wants these duties taken away from him. Safety,

9

however, is a part of these duties and cannot be surrendered to the safety director or a safety committee. Although it is one of the supervisor's jobs to prevent accidents, the safety director's job is to work with management to plan the company safety program and to help supervisors and management. The safety director gives the supervisor information about accident prevention. The wise supervisor works closely with him and with safety committees.

The supervisor's safety program

Assuming that the supervisor is given the necessary backing and that he is trained to recognize those factors in his workplace that have accident-injury potential, he can contribute to the safety program by training employees in proper work practices, by supervising and evaluating workers' performance with consideration given to unsafe acts and conditions, by adhering to all safety rules and regulations issued by the company, and by constantly monitoring the workplace for unsafe conditions and reporting them to management. The supervisor should investigate accidents occurring to workers, review all unsafe conditions with the workers, and ascertain whether personal protective equipment and safety devices are operational and are used and maintained as intended.

The supervisor should investigate every accident in his department, whether or not personal injury results. He needs to find the reasons for the accident, and take steps to prevent a recurrence. The supervisor should study his accident records to know what his most serious problems are, and where his department stands in relation to other departments in the company.

To be a good safety-minded supervisor as well as a good production-minded supervisor requires an understanding of good supervisory techniques. As a supervisor betters his department's safety performance, he frees himself of the annoyance of accidents and has more time for planning, improvement, and production.

This war against accidents continues as safety is built into the regular operation and management of production. Management increasingly recognizes the important work of supervisors in preventing injuries. Many companies have set their sights on injury-free work performance. Many have succeeded. The key to this success is the supervisor.

CHAPTER 2

Know Your Accident Problems

Where, how, and why do accidents occur? These three questions are asked by the progressive supervisor who makes accident prevention an important part of his job.

Sometimes the reasons seem obvious, but more frequently they are difficult to uncover. With few exceptions, however, they can be found.

The supervisor uses many methods to locate his trouble spots. Personal observation, for instance, often will uncover hazards that can be remedied directly. Examples are slippery spots on the floor, overflowing trash cans, and boards with nails protruding like rattlesnake fangs. Examples of easily found unsafe acts are failing to use guards or failing to wear personal protective equipment, or running on stairways.

Common sense alone can prevent many accidents. If, for example, a heavy machine is going to be installed, then the danger area should be blocked off. Hoists and supports should be double-checked to make sure the machine will not fall or slip and injure someone.

Job studies and inspections help locate trouble spots. Suggestions are given in later chapters of this Manual. NSC's *Accident Prevention*

11

Manual for Industrial Operations has about 30 chapters devoted to specific machine, process, and chemical hazards, in addition to many references and tables. The company safety professional should be an excellent source of help.

The most serious trouble spots are worked on first, for example, grounding portable electric tools, supplying lockouts for machines, or safeguarding nip points of belts and gears. Delays in correcting them can result in unnecessary injury or death. In one case, a young man was caught in the nip point of an unguarded belt and was killed. Instructions had been given three months earlier to install a protective barrier around the belt.

In many instances, a supervisor has the authority to make these corrections. In other cases, he is authorized to request that they be made by another department. Proposals for major safety improvements are usually submitted to an executive or to a central safety committee for consideration and approval.

It is essential that the supervisor follow up on all recommendations.

See and Know Job Hazards

An experienced supervisor should have little trouble in finding ways to eliminate accidents at the source, and in judging the effectiveness of these ways. Such a supervisor has already studied the equipment for the best ways of getting more production, of reducing down-time and of developing better quality. He has worked with his men to better their performance. The procedure for controlling hazards is similar.

At each step of an operation, the supervisor asks himself, "How can an accident occur here?" He usually finds the answer by observing the actions of the employee in relation to the material, the machine, or tool as he performs each step in the work.

For example, consider the operation of an ordinary abrasive wheel. By observing the operator at work, the supervisor can see these possible accidents, and suitable preventive measures, as indicated.

1. The employee's stance exposes him to the danger of being struck by flying pieces if the wheel should disintegrate. This could cause serious injury or death. Preventive measures include: using a

UNSAFE ACT   **UNSAFE CONDITION**

Both unsafe acts and unsafe conditions are involved in most accidents.

guard, "ring testing" and speed-testing wheels prior to use, selecting the right size wheel for the speed, and mounting the wheel properly.

2. The operator may strike his fingers or hand against the revolving wheel. Various precautions include: using a portable grinder on large work, filing small parts instead of grinding them, using tongs to hold parts against the wheel. When grinding, the operator should have the work rest securely clamped ⅛ in. from the abrasive wheel.

3. The operator's eyes may be struck by flying particles. The precaution is obvious: install a safety glass shield and require eye protection.

When studying a job for hazards, list each job step, describe likely accidents and how to prevent them. The safety professional, master mechanic, and others with specialized information can be of help here.

When approving or adopting changes in equipment, operating methods, or work practices, the supervisor should consider safety as well as efficiency. Safety should be designed in, insofar as possible. Hazards that cannot be "designed out" should be controlled by safeguarding. The employee can wear protective equipment, and should

13

follow established safe procedures. The supervisor must train men in use of the safety devices and in the proper procedures. He must follow up until the safe, correct methods are established.

Correct first things first

When establishing "a hazard hierarchy" (or "priority"), start with the sources of the most serious injuries, especially those that may lead to disaster—loss of many lives and heavy property damage. These sources include boilers and other high-pressure equipment, and equipment that uses or generates flammable liquids (like gasoline or naphtha), explosive gases (like acetylene or natural gas), or explosive dusts (like starch or other organic material), or equipment that can collapse from overloading or failing to properly secure.

Serious health hazards are other major targets. Silica dusts, lead, carbon monoxide, cotton dust, asbestos, and many other substances impair the health of workers and, unless controlled, may affect whole groups. Noise, lasers, and other forms of radiation are physical hazards that can adversely affect people.

Near misses or no-injury accidents, as well as injuries, should be investigated. A falling brick may merely glance off one man's arm, but crush the skull of another. The alert supervisor heeds a close call as a warning sign and takes prompt action because of what *might* have happened. He acts promptly to eliminate or control the hazard of the falling brick. He knows that if he does not, eventually the law of averages will catch up with him and another falling brick may cause serious injury.

Violations, too, should serve as warning signs to the alert supervisor. Repeated violation of safety rules, such as failing to wear adequate safety equipment or chance-taking and horseplay, eventually lead to accidents and resultant injury.

Certain equipment has a comparatively narrow hazard range—but these few hazards can produce severe injuries. Electric equipment and power presses have limited, but severe hazards—fatal shock and amputation, respectively. For this reason, special attention should be given to such clearly defined sources of serious injury.

Do not discount the possibility of a serious injury from a generally known hazard just because no accident has yet happened there. Minimizing the likelihood of a serious accident for this reason leads

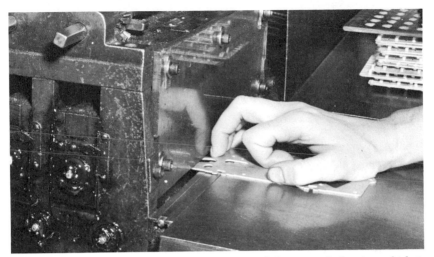

The inrunning rolls of this machine are protected by a simple barrier, which is adjustable to admit varying thicknesses of stock but not the operator's fingers. This inexpensive guard should prevent serious and costly injuries.

to the assumption that nothing need be done. One explosion that killed five persons, for example, happened with an organic dust that was known to be explosive. But because the department had operated for years with nothing happening, the hazard was disregarded.

The ordinary accident sources are readily identified by a simple check of equipment and materials used, but an awareness of both latent dangers as well as obvious dangers is necessary if hazards are to be controlled.

Review department experience

To locate trouble spots, review department accident experience. Take for example Table I, which shows injuries, by source, in two departments of a large company over a comparable time. In Department A, welding and cutting were the major problems; in Department B, hand tools, especially hammers, were involved most. Equipment and procedures should be studied.

In another firm, more than 55 percent of all injuries were "knife cuts." Investigation showed only a couple of cases were due to unsafe

TABLE I.

BREAKDOWN OF INJURY SOURCES

Source	Department A		Department B
Welding and cutting equipment and materials..		70	22
Plates and pipe handled manually.............		45	14
Hand tools			
Hammers............................	12		37
Wrenches...........................	6		7
Jacks..............................	5		0
Other..............................	10		9
Total hand tools.......................		33	53
Power tools			
Chipping hammers......................	10		2
Grinders...........................	2		2
Other..............................	3		4
Total power tools......................		15	8
Miscellaneous.............................		29	4
Total all..............................		192	101

handles or unsafe conditions. Unsafe practices—especially failure to use a knife holder—were the culprits. Safe use of knives was studied and safe practices established. Several short meetings of employees were devoted to instruction. One of the more skilled workers demonstrated proper techniques. Supervisors followed up by checking work practices. The result—knife accidents were almost eliminated. (A full discussion of knife safety is given in Chapter 11, Hand and Portable Power Tools.)

Systematic inspection

The supervisor should make sure that tools, machines, and other department equipment are maintained properly and are safe to use. To do this effectively, he uses systematic inspection procedures, and may delegate authority to others in his department.

16

SAFETY INSPECTION CHECK LIST

Plant or Department_____*SHIPPING*_____   Date_____4/1_____

This list is intended only as a reminder. Look for other unsafe acts and conditions, and then report them so that corrective action can be taken. Note particularly whether unsafe acts or conditions that have caused accidents have been corrected. Note also whether potential accident causes, marked "X" on previous inspection, have been corrected.

(V) indicates *Satisfactory*   (X) indicates *Unsatisfactory*

---

1. FIRE PROTECTION
Extinguishing equipment . . . . ☑
Standpipes, hoses, sprinkler heads
 and valves . . . . ☑
Exits, stairs and signs . . . . ☑
Storage of flammable material . . . . ☒

2. HOUSEKEEPING
Aisles, stairs and floors . . . . ☑
Storage and piling of material . . . . ☑
Wash and locker rooms . . . . ☒
Light and ventilation . . . . ☑
Disposal of waste . . . . ☒
Yards and parking lots . . . . ☑

3. TOOLS
Power tools, wiring . . . . ☑
Hand tools . . . . ☒
Use and storage of tools . . . . ☑

4. PERSONAL PROTECTIVE EQUIPMENT
Goggles or face shields . . . . ☒
Safety shoes . . . . ☒
Gloves . . . . ☒
Respirators or gas masks . . . . ☐
Protective clothing . . . . ☐

5. MATERIAL HANDLING EQUIPMENT
Power trucks, hand trucks . . . . ☑
Elevators . . . . ☐
Cranes and hoists . . . . ☐
Conveyors . . . . ☐
Cables, ropes, chains, slings . . . . ☐

6. BULLETIN BOARDS
Neat and attractive . . . . ☑
Display changed regularly . . . . ☑
Well illuminated . . . . ☐

7. MACHINERY
Point of operation guards . . . . ☐
Belts, pulleys, gears, shafts, etc. . . . . ☐
Oiling, cleaning and adjusting . . . . ☐
Maintenance and oil leakage . . . . ☑

8. PRESSURE EQUIPMENT
Steam equipment . . . . ☐
Air receivers and compressors . . . . ☐
Gas cylinders and hose . . . . ☐

9. UNSAFE PRACTICES
Excessive speed of vehicles . . . . ☑
Improper lifting . . . . ☒
Smoking in danger areas . . . . ☑
Horseplay . . . . ☑
Running in aisles or on stairs . . . . ☑
Improper use of air hoses . . . . ☑
Removing machine or other guards . . . . . ☐
Work on unguarded moving machinery . . . ☐

10. FIRST AID
First aid kits and rooms . . . . ☑
Stretchers and fire blankets . . . . ☐
Emergency showers . . . . ☐
All injuries reported . . . . ☑

11. MISCELLANEOUS
Acids and caustics . . . . ☐
New processes, chemicals and solvents . . ☐
Dusts, vapors, or fumes . . . . ☐
Ladders and scaffolds . . . . ☑

Signed_____*A. M. Baldwin*_____

USE REVERSE SIDE FOR DETAILED COMMENTS OR RECOMMENDATIONS

Systematic inspection is the basic tool for maintaining safe conditions and checking on unsafe practices. A detailed check list like this one helps the inspector find hazards before they can cause accidents.

Some companies use safety committees or special committees for periodic or special inspections. These inspections support—not supplant—the supervisor's responsibility for locating and correcting hazards.

17

Federal and state departments of labor compliance officers are empowered to enter, at any reasonable time and without delay, any establishment covered by safety and health regulations. An employer representative should accompany the compliance officer during the inspection.

Each operator may be required to inspect his work place and the equipment or machinery he uses. He does this each day and immediately reports defects that he is not authorized to correct. In departments having a number of similar machines, a specialist should make regular inspections and perform the necessary adjustments and repairs.

Mechanical and electrical departments of some companies are given the major responsibility for maintaining equipment in safe condition. A maintenance man may make the regular inspection. The mechanical department may follow a schedule for inspecting and maintaining hoisting cables, chains, and slings of overhead cranes, for example; and the electrical department may be responsible for all electrical parts of this equipment.

Toolroom men should inspect all hand tools to see that they are kept in safe condition. Some companies require portable electric tools to be turned in to the electrical department for a monthly check.

Inspection programs should be set up for new equipment, material, and processes. Nothing should be put into regular operation until it has been checked for hazards, its operation studied, additional safeguards installed (if necessary), and safety instructions developed. Good preventive maintenance produces substantial reductions in accidents and fewer production delays.

Crew leaders are often given the responsibility for inspecting equipment and for seeing that their men observe safe practices. The supervisor should make certain that all inspections by persons other than himself are up to his standards.

By systematic planning and organization, the supervisor can carry out his primary function—supervision. He spot checks periodically to make sure assignments are being carried out, that safety precautions are being observed, and that equipment is running efficiently and safely.

Serious injury has occurred because these procedures were not followed.

Legal requirements for safety

Safeguarding the health and safety of employees must conform to legal requirements. The Occupational Safety and Health Act (OSHAct) emphasizes the right of employees to a safe and healthful environment in a work-place free of recognized hazards. The Act imposes a legal liability on employers to conform to certain standards. These health and safety requirements usually deal with dangerous equipment and processes, but are often only minimum requirements. Legal requirements, moreover, cannot cover every specific operation or piece of equipment in every industry or company. Local requirements should also be checked.

Although a department or company safety program must conform to legal requirements, the steps taken by companies with the best safety records usually exceed these requirements. The goal is maximum performance, not minimum conformance.

19

OSHA poster, "Safety and Health Protection on the Job," right-center of bulletin board, must be posted conspicuously at every plant, job site, or other establishment.

Information sources

It takes at least about 500,000 man-hours of exposure to reveal definite accident trends or unusual sources of accidents. A 25-man department, therefore, would have to work about 8 to 10 years before its accident experience would reveal significant trouble spots. Yet this department may well have a number of minor and serious hazards that could result in accidents.

To prevent accidents from these hazards, the wise supervisor cannot rely solely on his own experience, regardless of how varied or lengthy it may be. He profits by the experience of other companies that continuously contribute to the body of accident prevention knowledge.

The sharp supervisor picks up this information in many ways. He hears about an accident in another department or another company that could have just as easily happened in his department. He then takes steps to make sure it will not happen to his men. He reads the

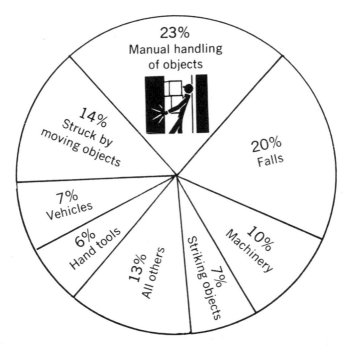

Handling of materials ranks high as a source of compensible work injuries. General statistics like these are helpful, but more specific information, based on actual experience, is eventually necessary.

Source: Reports from state labor departments as tabulated in *Accident Facts*.

accident prevention literature—books, magazines, newsletters, and pamphlets. (A listing is given in the "Useful References" section of this Manual.) He seeks expert help—from the technical literature and from experts—for special problems.

For example, the supervisor can find in National Safety Council's annual *Accident Facts* some of the general sources of accidents in his industry, as well as in all industry. These statistics show that year after year, about 20 to 25 percent of all compensable injuries involve handling of materials. Even though the supervisor lacks other information on the subject, this one statistic should alert him to the hazards of materials handling. Many insurance companies also have statistical data that may be of value here.

When the supervisor has a special problem—such as presented

by toxic solvents, flammable liquids, electric equipment or machine guarding—he can uncover a wealth of information. If, for instance, highly toxic carbon tetrachloride must be used in his department, he can quickly read up on the subject in a National Safety Council Industrial Data Sheet.

Not only should the National Safety Council be consulted, but also the American National Standards Institute, the U.S. Bureau of Mines. Underwriters' Laboratories, Inc., insurance companies, trade associations, professional societies, state and local health authorities, and other organizations with specialized information. The chapter "Sources of Help" in NSC's *Accident Prevention Manual for Industrial Operations* lists more than 100 sources of information in United States, Canada, and other countries of the world.

The company safety professional can be of great assistance in developing information. Often he refers to appropriate literature or directly consults with experts.

Although the National Safety Council does no inspection or field consultation, its staff and its hundreds of volunteer officers stand ready to offer advice without obligation. Most firms, too, will exchange information about accident problems. Insurance company safety engineers can also help the supervisor.

The supervisor's knowledge of his department's operations and his company safety professional's experience are an excellent combination for judging accident problems and planning safety work. In most cases, the safety professional has departmental reports covering accidents over a period of years. Analysis of the circumstances will disclose useful information about the principal accident sources.

Accidents that require only first aid treatment should be included in these analyses because the circumstances might have resulted in more serious injuries. Sources of frequent injuries continue to be important problems year after year, until they are dealt with effectively.

Elements of an Accident

The supervisor who blames all accidents on "carelessness" of his men is no better off than the fellow who tries to start a car by kicking the tires. The mechanic who understands how the parts of a car work has the basic information for locating trouble and correcting it. Similarly, the supervisor who knows the elements of an accident and how they contribute to an injury has a tool for securing the right facts upon which to base corrective action. He also has a tool for effectively planning and performing all his safety duties.

No mystery should surround an accident. It can be taken apart, like an auto, to find out what went wrong. Each contributing element should be analyzed, and then reported so that the information can be used for preventing other accidents.

Here are the five elements:

1. **The agency or source of the accident** (the item involved, such as a hammer, a piece of sheet metal, or a lift truck). Also the part of the tool, material, or equipment involved, in order to pinpoint the corrective action.

2. **The type of accident** or the manner in which the person was injured (such as by falling, by being struck by an object, or by getting caught in or between moving equipment).

3. **The unsafe condition** of the tool, material, or machine (such as a broken handle of a hammer, unguarded gears of a machine, or worn brakes of a lift truck).

4. **The unsafe act** or unsafe practice of the person (such as working near the moving parts of a machine without first stopping it and locking it out, lifting with the back muscles instead of the leg muscles, or removing a guard).

5. **The personal factor** or reason for the person's unsafe action or practice (such as lack of knowledge of the safe practice, disregard of instructions, physical handicap, or emotional upset).

Here is an example of how a supervisor analyzed one accident.

The facts: Two men were working near the end of a scaffold, about 15 feet off the ground. John told Henry to move out of the way so that he could more easily saw a board. Henry backed up without looking, fell off the end of the scaffold, and broke his left ankle when he hit the ground. The ends of the scaffold were not enclosed.

In investigating and reporting the accident, the supervisor looked for the facts. He tried to report them without bias, without blaming either man. This attitude is essential to effective reporting.

The supervisor considered all five accident elements:

1. The scaffold was the type of equipment involved in the accident.
2. A fall was the type of accident (the way in which Henry was injured).
3. The lack of a guard rail was the unsafe condition.
4. Backing up without looking was the unsafe act that precipitated the fall.
5. The reason for the unsafe act was that Henry forgot momentarily the danger of the open end of the scaffold.

Analysis of this accident clearly shows that lack of the guard rail was the major cause of the accident, rather than Henry's action. The man would not have fallen, regardless of his action, if there had been guard rails and toeboards as required by OSH Standards for scaffolds more than 10 feet above the ground or floor.

Unsafe conditions and unsafe acts

Generally, elimination of an unsafe condition is the most effective way to prevent a similar accident. When analyzing an accident, always look first for an unsafe condition.

A common example of safeguarding against a possible unsafe condition is the three-wire cord and three-prong plug which permits automatic grounding of electric equipment and eliminates the worry about getting a man who is using an electric drill, for example, to connect a separate ground wire to a water pipe. The three-prong setup, if properly grounded and maintained, is safety-engineered to prevent the unsafe act.

The unsafe practice (or act) becomes the principal contributing factor only when there is no unsafe condition, or when no safer method of doing the job can be devised.

For years it was thought that unsafe acts were responsible for about 85 percent of all disabling injuries, and unsafe conditions accounted for the remaining 15 percent. Now, safety professionals realize there seldom is a clear-cut single cause. Actually, both unsafe conditions and unsafe acts are involved in at least 4 out of 5 accidents.

An example of an accident involving only an unsafe practice would be when two men are carrying a length of pipe. One man drops his end without warning the other, who loses his grip and drops his

An unsafe act can be as simple as a man over-reaching and twisting while handling material.

end on his legs or feet. It may be that more than one unsafe act was involved. The men may not have been trained to carry the pipe on the same side of their bodies, or one of the men did not watch where he was going.

Painstaking investigation may be necessary to uncover all unsafe conditions or practices. Careful study of the findings may be necessary to find the most effective preventive measures. An inadequate remedy resulting from incomplete facts about an accident may be worse than no remedy at all.

Accident investigation

Accidents should be reported at once and investigated immediately. Near-miss accidents and accidents resulting only in property damage should also be investigated.

When a man is injured, the supervisor must take emergency

Accident investigations are conducted to prevent other accidents, not to fix blame.

action to have first aid administered, to obtain professional medical attention at once, and to protect fellow employees and equipment. He must then begin to investigate the circumstances of the accident.

The supervisor should use a form to report the injuries. This report form should elicit significant information about an injury in order to help the supervisor get to the real causes of the situation—those factors which, if eliminated, will prevent a recurrence.

These procedures have been found particularly effective when investigating accidents:

1. Go to the scene of the accident promptly.
2. Talk with the injured person, if possible. Talk with witnesses. Stress getting the facts, not placing the responsibility or blame.
3. Listen for clues in the conversations about you. Unsolicited comments often have merit.
4. Encourage people to give their ideas for preventing the accident.
5. Study possible causes—both unsafe conditions and unsafe practices.
6. Confer with interested persons about possible solutions. The problem may have been solved by someone else.

7. Write up a report, using a printed form which allows a narrative description. (One is shown on the next page.)
8. Follow up to make sure conditions are corrected. If they cannot be corrected immediately, report this to all concerned.
9. Publicize any corrective action taken so that all may benefit from the experience.

When following this procedure, keep these two basic points in mind:
- That most accidents involve both unsafe conditions and unsafe acts, and
- That the purpose of accident investigation is prevention of future accidents and not fixing of blame.

The supervisor should, therefore, look for all causes—both hazardous conditions and human failures. If he is to dig deep enough to be effective, the supervisor must not jump to conclusions. He must also realize that people will suspect the value of an analysis if it places more than a fair share of blame on them.

Considering the second point, the supervisor must accept this assignment of controlling unsafe acts and unsafe conditions without trying to whitewash himself. All too frequently the excuses, "The fellow failed to follow my instructions!" or "The man was just plain careless," are given, when actually the supervisor really knows he failed to instruct properly in the first place.

Measuring Safety Performance

How does a supervisor know if he is getting results? How does he compare his safety performance with that of other supervisors? How does he measure his own progress? Or failure?

The answer—standard reporting procedures. There are two methods for measuring safety performance.

1. Accident cost—medical cost, compensation or insurance cost, production loss due to accidents.

2. Recordable injuries and illnesses. (Discussed later in this chapter.)

By using these yardsticks, the overall accident situation in a department may be gaged by comparison with the injury experience of

27

2—Know Your Accident Problems

Supervisor accident report form, available from National Safety Council, encourages accurate and complete reports, invaluable for pin-pointing sources of injury (illness) and unsafe practices so that action can be taken.

28

22. Personal protective equipment required? (Protective glasses, safety shoes, safety hat, safety belt) _____

    Was injured using required equipment? _____

23. What can be done to prevent a recurrence of this type of accident?
    (Modification of machine; mechanical guards; correct environment; training)

_____

_____

24. Detailed narrative description (How did accident occur; why; objects, equipment, tools used, circumstance, assigned duties.

    Be specific) _____

_____

_____

_____

(Use additional sheets, as required)

25. Witnesses to accident _____

_____

Date prepared _____   Signature of Foreman / Supervisor _____

     Department _____

SUPERINTENDENT'S APPRAISAL AND RECOMMENDATION

a. In your opinion what action on the part of injured (or ill) person or others contributed to this accident?

_____

_____

b. Your recommendation _____

_____

Date _____   Signature of Superintendent _____

FOR SAFETY OFFICE USE ONLY

Temporary Total ☐	Permanent Partial ☐	Death or Permanent Total ☐
Started losing time _____	Part of Body _____	
Returned to work _____	Per cent loss or loss of use _____	
Time charge _____	Time charge _____	Time charge: 6,000 days

Compensation $_____ Medical $_____ Other $_____ Total $_____

Name and address of hospital _____   Name and address of physician _____

Reverse side of report. Central portion is filled in by higher-level management. Bottom portion of form, filled in by safety department (as the facts become known), contains data for computing injury (occupational illness) rates and costs.

similar operations in other companies. A favorable comparison indicates a good standard of safety. The best records provide a goal toward which employees of departments with poorer records can strive.

Accident costs

Because they recognize accident prevention as a significant part of the supervisor's job, some companies set up their accounting procedures to record medical and compensation expenses by department. Some even list supervisor's names, with accident costs incurred by their respective departments, in monthly or annual reports to top management.

So that valid comparisons of accident cost can be made between departments (and companies and industries), the exposures must be equalized. Standard cost factors are usually based on cost per 1,000 man-hours worked, or per $100 of payroll.

Departmental accident cost may be classified by source of injury, such as portable electric tools or handling materials. This type of classification system permits supervisors to find their most expensive trouble spots.

Accident cost may be classified another way—by listing as either insured or uninsured.

- *Insured costs* include compensation insurance premiums and, in some cases, medical expenses. These are easy to obtain.
- *Uninsured costs* are more difficult to determine, but are estimated by the National Safety Council to be about equal to (and sometimes higher than) the insured or visible costs. Those costs that can be determined should be placed on the supervisor's accident cost report. These items can usually be costed: production loss, time lost by supervisors, time lost by co-workers of injured employee, losses from damaged materials or machines, and other losses.

Let's discuss each one.

1. Production losses come from partial or complete shutdown due to:
   a) Damage of material, machines, or production area.
   b) Emotional upset and lowered employee morale which tends to lower production.

30

c) Increased tension resulting in an increase of materials spoiled
   and pieces rejected.
 d) Replacement employee(s) who produce less while being trained
   to fill injured worker's job.

2. Time lost by supervisors can be due to:
 a) Assisting injured employee.
 b) Investigating the accident.
 c) Preparing accident reports.
 d) Hiring and training new employee(s).
 e) Attending hearings conducted by the state.

3. Time lost by co-workers of injured employees:
 a) In aiding injured person(s).
 b) Because of curiosity.
 c) Because of sympathy.

4. Losses from damaged material or machines due to:
 a) Cost of repair of building, machines, or tools.
 b) Cost of damaged or spoiled pieces.
 c) Losses in wages paid to injured employee when he returned,
   even though his services for a time are restricted.

5. Other losses:
 a) Loss of business because of late deliveries.
 b) Loss of good will and prestige.
 c) Grievances, cost of impairment of employer-employee relationship.

   Those costs that can be determined should be entered on the
Supervisor's Accident Cost Report. These items can usually be costly
—wages lost to nonproductivity of non-injured workers, damage to
material or equipment, wages paid for lost time by the injured (other
than Workmen's Compensation), overtime work when necessitated by
the accident, wages paid to supervisor and others for time spent on
activities made necessary by the accident.

   From a review of the Department Supervisor's Accident Cost Report, it would be a relatively easy matter for us to compare the direct
or insured cost and the indirect or uninsured cost of an accident. If,
for example, the injured worker received a minor injury, which required treatment by a doctor at a cost of $15.00, we would know that

DEPARTMENT SUPERVISOR'S ACCIDENT COST REPORT

Injury Accident __Minor injury__

No-Injury Accident_____

Date __Nov. 30, 19--__ Name of injured worker____John Smith_____

1. How many other workers (not injured) lost time because they were talking, watching, helping at accident? __3__

   About how much time did most of them lose? _____ hours __15__ minutes **each**

2. How many other workers (not injured) lost time because they lacked equipment damaged in the accident or because they needed the output or aid of the injured worker? __2__

   About how much time did most of them lose? _____ hours __40__ minutes **each**

3. Describe the damage to material or equipment __Unsafe sling allowed crate to__ __fall while being hoisted. Crate and separators broke.__

   Estimate the cost of repair or replacement of above material or equipment $__175.00__

4. How much time did injured worker lose on day of injury for which he was paid? __6 hours__ hours _____ minutes

5. If operations or machines were made idle: Will overtime work probably be necessary to make up lost production? Yes ☒, No ☐. Will it be impossible to make up loss of use of machines or equipment? Yes ☐, No ☒.

   Demurrage or other special non-wage costs due to stopping an operation $__75.00__

6. How much of supervisor's time was used assisting, investigating, reporting, assigning work, training or instructing a substitute, or making other adjustments __4½__ hours _____ minutes.

   Name of supervisor_____

Fill in and send to the safety department not later than day after accident.

Published by National Safety Council
425 North Michigan Avenue
Chicago, Illinois 60611

Form IS7   Printed in U S A   Stock No. 129.27

Accident cost report form (8½ × 11 in.) should be prepared by the department supervisor as soon after the accident as information becomes available on the amount of time lost by all persons and the extent of damage to product and equipment.

32

the direct or insured cost was $15.00. Making the assumption that the employees are paid at an hourly rate of $4.00 an hour and supervisors $6.00 an hour, we can compute the following cost. Three workers lost time because they were either talking, watching, or helping at the scene of the accident. Each of these workers lost 15 minutes. Two other workers lost time because they lacked equipment damaged in the accident or because they needed the output or help of the injured worker. Both of these workers lost approximately forty minutes each. The injured worker lost a total of six hours; he was paid his full wages during this period. The supervisor's time used in assisting, investigating, reporting, assigning work, training or instructing a substitute or making adjustments, amounted to 4½ hours. Therefore, a labor cost of $59.33 is developed. The estimated cost of repair or replacement of the damaged equipment is approximately $175.00. Overtime work is necessary to make up for the lost production and will cost $75.00. Therefore, the direct and indirect cost of this minor accident is:

DIRECT COST

Payment to the doctor	$ 15.00

INDIRECT COST

Labor cost		59.33
Repair or replacement		175.00
Overtime costs		75.00
Total indirect (or hidden) costs		$309.33
Direct cost of the above	$ 15.00	
Hidden cost of the above	309.33	
TOTAL COST	$324.33	

Recordable injuries and illnesses

The Williams-Steiger Occupational Safety and Health Act requires most employers to keep records of the occupational injuries and illnesses of their employees. Every employer covered by this act must keep these records up-to-date, have them available for inspection by government representatives upon request, and post a summary of all recorded cases at the end of the year.

Injuries and illnesses must be recorded separately with seven illness categories provided. Cases that must be recorded are: fatali-

LOG OF OCCUPATIONAL INJURIES AND ILLNESSES

OSHA NO. 100

Case or file no.	Date of injury or initial diagnosis of illness. If diagnosis of illness was made after first day of absence enter first day of absence. (mo./day/yr.)	Employee's name (First name or initial, middle initial, last name.)	Occupation of employee (enter regular job title, not activity he was performing when injured.)	Department in which the employee is regularly employed.
1	2	3	4	5

Company Name --

Establishment Name---

Establishment Location---

NOTE: This is not a report form.

Keep it in the establishment for 5 years.

Log of Occupational Injuries and Illnesses, Form OSHA-100, is used to record injuries or illnesses which result in fatalities, lost workdays, or cases that result in transfer to another job or termination of employment, or require medical treatment, or involve loss of consciousness or restriction of work or motion. In

CORDABLE CASES: You are required to enter on this log information about each occupational **death**; each occupational **illness**; and each occupational **injury** which results in one or more of the following: loss of consciousness, restriction of work or motion, transfer to another job, or medical treatment (other than first aid). More complete definitions appear on the other side of this form.

Form Approved
OMB NUMBER

DESCRIPTION OF INJURY OR ILLNESS		EXTENT OF AND OUTCOME OF INJURY OR ILLNESS						Transfers and Terminations
Nature of injury or illness and part(s) of body affected (Typical entries for this column might be: Amputation of 1st joint right forefinger Strain of lower back Contact dermatitis on both hands Electrocution–body)	Injury or Illness Code See codes at bottom of page.	Fatalities *(deaths)*	Lost Workday Cases				Nonfatal Cases Without Lost Workdays	(If the entry in columns 9 or 10 represented a permanent transfer or termination, enter a check in this column.)
		Enter date of death (mo./day/yr.)	If case involved lost work-days, enter a check in this column.	Lost Workdays			If no entry was made in columns 8 or 9 but the case is recordable *(see definition above),* enter a check in this column.	
				Enter days *away from work* due to injury or illness.	Enter days of RESTRICTED ACTIVITY due to injury or illness.			
6	7	8	9	9A	9B		10	11

Injury Code

10　All occupational injuries

Illness Codes

21　Occupational skin diseases or disorders
22　Dust diseases of the lungs (pneumoconioses)
23　Respiratory conditions due to toxic agents
24　Poisoning (Systemic effects of toxic materials)

25　Disorders due to physical agents (other than toxic materials)
26　Disorders due to repeated trauma
29　All other occupational illnesses

addition to this form, there are a supplementary record of occupational injuries and illnesses, and an annual summary. Complete instructions for using the Log are on its reverse side. The form measures 9½ by 15 inches. These forms are subject to change. Be sure to check your OSHA Area Office for the latest forms.

35

Case No. *78*   Date *6/8*

First Aid Report

Name *John Smith* Check No. *14A* Department *Stock*
Male ☒ Female ☐ Occupation *Trucker* Foreman *J. Riddle*

Date of Injury *6/7* Time *2* ~~a.m.~~ p.m. Date of First Treatment *6/7* Time *2* a.m. ~~p.m.~~

Nature of Injury *Splinter right Thumb*

Sent: Back to Work ☒   Doctor ☐   Home ☐   Hospital ☐

Estimated Disability *O* days

Injured's Description of Accident *Moving splintered pallet*

Signed *J. Weir R.N.*
First Aid

Issued by National Safety Council, Inc., 425 N. Michigan, Chicago, Ill. 60611
STOCK No. 129.26—1S—6—Rep. 500P.—5-65-03

First aid reports, too, may give good clues to accident causes. For example, a number of "splinters in hand" cases might indicate the need for gloves, or for better maintenance of splintered or rough pallets and other equipment or in-process materials.

ties, lost work day cases, and non-fatal cases without lost work-days but which require medical treatment. Medical treatment includes treatment administered by a physician or by registered professional personnel under the standing orders of a physician. Medical treatment does not include first aid treatment even if the first aid treatment is provided by a physician. First aid treatment is one-time treatment and subsequent observation of minor scratches, cuts, burns, splinters, and so forth, which do not ordinarily require medical care.

Even though the first aid treatment cases are not required to be recorded or reported, they are important because they could have been more serious injuries. A record should be kept of them, not only for accident prevention purposes, but in case they later develop into compensation cases.

It is usually up to the doctor only to decide when the injured person shall remain away from work and when he shall return to work, or whether he is capable of performing the duties of a regular job.

Note that only cases that are work-related are included in the accident experience of a department. An accident to a man driving his car to work is an off-the-job (OTJ) accident; if he is driving a car as part of his job and has an accident resulting in an injury to himself, this is recordable.

Recordkeeping requirements are under constant review by OSHA. Be sure to check your Department of Labor area office for the latest requirements and forms.

Incidence rates

You can use the number of recordable injuries and illnesses and the number of first aid treatment cases to compute the incidence rate for your department. The incidence rate is the number of injuries for every 200,000 hours worked. By using the incidence rate, units of any size—departments, companies, and industries—can be compared.

Here's an example of how the incidence rate is computed for a unit of 300 employees with 12 recordable injuries.

300 employees working 50 weeks a year

50 weeks × 40 hours/week × 300 employees = 600,000 hours worked during the year.

$$\frac{12 \text{ injuries} \times 200,000}{600,000 \text{ hours}} = 4.00 \text{ incidence rate}$$

If operations and hazards differ greatly between departments, incidence rates cannot be a fair gage of each supervisor's effort. Even if departments are comparable, chance variation may affect the validity of comparisons based on experience of only two years.

The comparative percentage improvement of each department over a three- to five-year period would more accurately show up trouble spots or reflect progress.

Accident prevention is a year 'round job. The goal of safety work is to achieve longer and longer periods without injuries. Progress is gradual. Ultimately, however, incidence rates may become so low that changes have little significance and records can then be measured in number of man-hours worked since the last recordable injury.

The supervisor who works to know his accident problems is in a good position to evaluate them, correct the most troublesome hazards first, get measurable results in fewer injuries and lower costs, and get the credit he deserves.

CHAPTER 3

Human Relations

To produce goods and services is the supervisor's task. But unless he can get his men and women to work safely and efficiently, production will suffer. The attitudes and thus the performance of a group of employees is directly influenced by the supervisor. To run a safe, efficient department, the supervisor must know about the needs of his people and what they respond to. He must know what motivates his employees.

Background of Motivation

During the 1960's, it became popular for supervisors to attend sessions on human relations. Much human relations training is centered around motivation.* Most of these theories have been discovered by experimental psychologists and have been tested in experiments with both animals and humans, and have been substantially confirmed. Unfortunately, there are very many theories.

The classical approach to motivation is centered around changing the environment in which a person works—it is much easier to change

* "One More Time: How Do You Motivate Employees?" Frederick Herzberg. *Harvard Business Review,* January-February 1968, pp. 53–59.

Good personal relations help people work better.

the environment to conform to the employee's motivational pattern than it is to change the employee. For example, it is much more effective to put a safeguard on a machine than it is to try to keep employees safe by telling them to be safe, or to be careful.

Principles of motivation

People are motivated to act in order to satisfy their needs. These needs fall into two categories, physiological needs and social needs. Physiological needs are primarily those of food and shelter. In our society these needs are taken care of by paying employees an adequate salary. Most theories say that once these basic physiological needs are satisfied, the social needs become more powerful. This is why so much stress is placed upon the social needs or good human relations as motivating factors in industrial production. Social needs such as (*a*) knowing work has value, (*b*) appreciation, (*c*) security, and (*d*) feeling of belonging are among the strongest job motivators.

People will not do their best unless the work environment satisfies these social needs.

The supervisor should encourage beginners and let them know how well they are doing.

Knowing that work has value

People want to know that their job has value—that it is essential. People want a job that provides an opportunity for personal satisfaction or growth. They are more likely to have that feeling if they know where their job fits into the total pattern of the company. Often the job is so small a part of the whole production pattern that it is hard to see where it fits in.

The supervisor can help by explaining the company's overall objectives and by showing his people where their jobs are valuable to the company. The vital role that safety plays in the entire production picture must be emphasized.

A supervisor might think that everyone would realize that he would not be doing work unless it had value, but this is not always true. When anyone does the same thing over and over again—especially when this task does not produce a complete product—he can only keep a sense of the importance of his job when he has a clear idea of the final product of which his work is a part.

The person usually feels a sense of accomplishment just by learning a job; but after years of doing the same thing, this feeling of accomplishment needs to be reinforced. The person develops a feeling that his efforts are swallowed up in the vast quantities of similar work turned out by others.

The supervisor, in these cases, can help by having friendly discussions centered around working conditions, growth possibilities, or the work itself.

Knowing that work is appreciated

People want to know their efforts are appreciated. To belittle a man's work is to belittle the man himself. Praise works the other way.

The supervisor should give credit when it is due—but with caution. Giving a man a slap on the back and saying "Well done!" only means something if it is done sincerely, and is not just a mechanical gesture. Praise must also be only for good work. Praise for work that the employee knows is poor will not only make the supervisor look foolish, but will also reduce the value of future praise, even when it is justified.

A beginner's work may not be up to the quality standards of an "old timer," but the beginner should be encouraged by telling him that he is making progress (providing that he really is). He should be given further instruction to help him attain acceptable work.

Although the experienced worker generally knows when his work is good, he will be pleased, even so, to know that his boss knows it too. Perhaps the best general rule is for the supervisor to maintain a positive interest in the workers of his department. If he takes such an interest, he will know what job each worker does, praise those who do well, and encourage those who are still learning or do not do quite as well as they should.

Praising a worker for following safe practices will help impress upon him how important his supervisor considers safety to be, as well as encourage him to continue to do his job safely.

If people know their supervisor is not just looking for faults, but also will give them credit for good work, they will, in turn, have a better attitude toward their work. No one is likely to try to do more than the absolute minimum if he feels that what he does is not appreciated or recognized.

A supervisor well knows his own reactions when he himself is

Supervisors can promote a feeling of belonging by means of friendly discussions.

roughly handled by his own supervisor. Often it makes little difference whether it is his decision, or his general competence, that is questioned. Justified or not, criticism is a threat to his self-respect. A worker feels the same way.

The supervisor who attempts to maintain standards by using a "meet-them-or-else" attitude is more than likely to find his men busily looking for ways to get around meeting those standards.

Using explanation and encouragement has been found to be more effective than laying down the law and demanding conformity, which usually meets with passive resistance.

Having a sense of security

People want to feel secure in their jobs. Some think this means that people want good pay and steady work. They do. But they also want to know what is going on and what to expect in the future. People want to know what they are expected to do, and how to do it safely and efficiently.

42

Supervisors are responsible, therefore, to see their workers are thoroughly trained in their assignments. This is essential to safe, efficient work, and contributes to the worker's feeling of security. Whether the supervisor does the training himself or assigns another to do it, the ultimate responsibility is his.

Injuries, work spoilage, and misused or broken equipment are almost certain to result when a worker has been inadequately trained. Such casualties and production delays reflect on the individual, supervisor and department—and reduce the security of all.

If a man has had poor training and feels as though he has never mastered his work, he will continually be uncertain and hesitant. He may develop a sense of defeat and become uninterested in his job. This will lead to dissatisfaction and poor work.

Training should include how to avoid getting hurt. A man should know that his company wants to protect him against injury. He should know what he can do to protect himself. Accident prevention should be part of a man's training. Special emphasis on safety is desirable because it creates an interest in safe practices for their own sake.

Belonging to the work group

People want to belong to their work group. The good supervisor will do his best to create and maintain a feeling of group solidarity and friendliness.

He can watch for opportunities to talk with his group and, in talking to them, arouse group interest. Using such words as "we," "our department," and "our safety record" encourages group interest and a team spirit.

The supervisor, when seeing a man is left out or has excluded himself from the group, can look for chances to include him.

Calling men by their first names aids the group feeling. The use of name badges helps in large departments.

A small clique, which excludes other members of the work group, is harmful. The best way to combat this is to give all members the feeling of satisfaction that they belong to the larger group.

The group feeling promotes good performance and safe practices. Once the group accepts them, members discipline one another. Often, the group standards are higher than the standards of most of the individuals by themselves. The supervisor can appeal to the pride of his people, in their safety record, and in their consistent safe practices.

One thing the supervisor should not do, if he wants to preserve a healthy group feeling, is to tell his people, "You fellows have got to . . ." or "Now listen, I want you guys to . . ." The supervisor needs to be an accepted member of the group and can soon exclude himself by such an attitude.

The supervisor must appeal to the self-interest of the group. When he appeals to each member to follow a line of action or a pattern of conduct, he must make sure that such action or conduct is for the member's and the group's own good.

Most people do not want to stand out as exceptions. They want to be liked by others so they try to think and act as they think others do. This does not mean they are forced to conform, they usually conform because they have similar backgrounds to the others and actually believe in the same standards.

Neither does this mean that individuals *never* want to be considered exceptional—never want to be singled out as having an unusual ability. A man may be proud of his brute strength or his ability to produce more than anyone else in the department and he may want others to recognize this. But even these men will generally exercise their special talents only within limits allowed by their group.

The man who consistently tries to prove his abilities, however, probably does so because he feels he is not a full member of the group. Some may consider lack of acceptance to be a real handicap. Others find this less important. By and large, though, people need to be part of their group. This desire can affect their actions.

The supervisor uses this desire for group membership when training the new workers. If the group works safely and efficiently, the new member will be likely to work this way too. The fact that the new man wants to be a member of this group will be a powerful incentive to work safely and efficiently.

The Supervisor as a Leader

The supervisor's big job is to provide effective leadership. With a people traditionally used to making decisions for themselves (as people do in the open and partially abstract society of the United States and Canada), leadership works best when it enlists the cooperation of those who are being led. Much research has been done over the past 30 years to determine the qualities of a good leader.

Here are some of the more significant ones together with a few comments. The qualities listed are what would be expected and desired of a good leader.

1. He should realize and show that many of the ideas he uses come from the people in his group. In this connection, many studies show that leaders do not originate all of the plans; in fact, with many good leaders, the majority of the ideas do originate in the group. It is common failure of new leaders to feel that they lose face if they accept suggestions from subordinates. On the other hand, experienced leaders know the value of having the group feel that they are participating and that goals set up reflect their own ideas and contributions. In some instances, the smart leader will inspire the group or a group member to suggest things that he already has in mind.

2. He shows respect for employees both as individuals and as a group. He should be sensitive to, and understanding of their needs.

3. The group wants the leader to be the kind they would have selected themselves, even if he be appointed by the company. They want the true, natural leader.

4. The wise leader should be, and is, satisfied to know he is right. He never feels that he has to prove that others are wrong.

5. He should act as a buffer between employee and higher management by not passing on the pressures put on him to those who work for him. This includes taking personal responsibility for the directions and orders he is required to give even when they are not what he, personally, would like to do or would have chosen. He never says, "The top brass says we have to do this," because it gives others the feeling that he is shirking his responsibility.

6. He leaves his personal problems and feelings out of his relationships with the group and its members by showing a calm, understanding approach at all times. A lack of emotional balance stirs up tension and stress in the group, particularly when he is grumpy, irritable, and irascible one day and pleasant, happy, and agreeable the next.

7. He is fair and square at all times and plays no favorites—always in keeping with good leadership.

8. He lives up to what he says by setting the right example. The old cliché, "Don't do what I do; do what I say," just does not work in on-the-job human relations.

9. He is available and understanding when people come in for help. He avoids the impression that he never has time to help or to listen to problems. When a supervisor is really under pressure and just can't take the time at the moment, there is nothing wrong in his saying so and telling people he will be with them just as soon as he can, provided he actually does live up to his promise.

10. It is important to be able to be accepted as one of the group, but it is just as important not to go so far that status as a leader is destroyed. Unfortunately, some leaders have tried so hard to be "one of the boys," particularly in off-the-job get-togethers, that they have crossed that indefinable (but very real) line that separates leaders from those they lead. People realize that this line exists and that it has to be maintained; when it is not, status as a leader is destroyed.

11. The job climate must be such that there is a minimum of conflict and confusion which would generate fear and anxiety in group members with resultant inefficiency in all operations.

Being a good leader, able to supply the emotional, mental, and physical needs of those dependent upon us, is humanity at its finest.

Preparing the Worker for Changes

To the degree a job changes, it becomes new. A person must be kept abreast of these changes if he is to remain qualified.

A person must be warned of job changes, or else he will be uneasy about them even if he wants the changes. He is usually a little uncertain about just what will be expected of him in the new situation and he will try to visualize himself in the new environment.

If a person is not warned until after the change is over, he has no time to adjust to the idea. If the change remains unexplained, a person worries about the personal meaning of this.

Briefing employees on work loads and planned changes ups production and morale.

Because everyone usually identifies himself with his work, he goes on the defensive when his work is found at fault. He feels it is not just his work, but his own competence that is being questioned. Similarly when job changes are made. It is not just the job that changes, but the man's work habits and his conception of himself as a worker.

The change usually requires something new of him. His old skills may no longer be needed. This may frustrate him. He may also feel unimportant because he knew nothing in advance about what he was going to be doing.

Those men who try to retain their feelings of security and job competence may actively resist change. Providing advance knowledge of the change helps undercut this resistance. Letting the men participate in the decisions leading up to the change will be even more effective. Employees who feel they have a part in the decision take more interest in it. They feel worthwhile and actually increase their sense of security.

The degree to which workers can participate varies greatly between companies, but it should still be there. Everyone benefits from free exchange of ideas. Workers know their knowledge is appreciated and their opinions respected. Management gains in securing both the knowledge and the cooperation of the employees.

It is important that management make clear from the start how much consideration will be given these suggestions. If knowledge and opinion are sought, they should be taken into consideration. This does not mean whatever is suggested must be done. There can be honest differences of opinion. But, because decisions must be made, it may be necessary that they run counter to someone's opinion. In such instances, the reasons for the decision should be explained.

Men must be given credit for their suggestions that are accepted. If people are not given credit, their desire to think constructively about their jobs becomes dampened. No company can afford this.

Workers with Special Problems

In all but exceptional cases, workers have the ability to do their jobs well and safely. They are reasonable adults who expect to be required to meet certain standards and who will want to do so.

Most workers are normal. Job interviewing and screening sees to this. They want security, recognition, and a sense of belonging to their group. If given a chance, they seek satisfaction of these needs through acceptable activities. Some, however, may have dispositions that interfere with the primary purpose of the department—the production of goods and services. What does the supervisor do when "normal" problem-solving methods fail? If he finds someone who does not seem normal, he can refer the problem to his personnel department or to his own superior.

Unfortunately, recognition that people who bring their personal problems to work with them are less safe and efficient has led, sometimes, to overemphasis on the abnormal personality. Perhaps the reason is that it is easier for the psychologist to describe abnormal behavior than to describe normal behavior.

The supervisor needs to know effective methods of training and motivating normal people to do their work well—not methods of treating mentally sick people.

To help his people work better, the supervisor must be on good personal relations with them and often acquire a knowledge of their personal problems. Although he must know a lot about people if he is to supervise them competently, it is not his responsibility to try to straighten out their private affairs or personality difficulties. Attempts to do this could do more harm than good.

Temporary problems

Proper personnel placement tries to make sure that a man either knows his job or has the ability to learn it. Even proper placement, however, cannot eliminate some crisis in a man's private life that makes him temporarily inattentive to his job. If a supervisor is friendly and approachable, his men often tell him that they will be unable to do exacting or hazardous work for a few days. Physical work, if not hazardous, is often just what a man needs to temporarily escape some personal problem and later to better cope with it. A badly disturbed man, of course, should not be allowed to work on a job where he may injure himself or others. A supervisor can often serve this fellow best by referring him to the personnel department.

In contrast to normal workers who are temporarily disturbed by personal difficulties, there may be others who have abnormal personalities and so constitute relatively long-term problems. How does one recognize such workers? People who are under severe emotional strain may not show visible signs of it for some time.

A signal that a worker needs help is when he shows signs of undue uncertainty or anger or fear in commonplace situations during work hours. If a top workman starts turning out shoddy work or getting easily upset, he may need help. Antisocial behavior is another sign. A person who continually disrupts the work group may need counseling. In general, these people should be referred to the personnel department.

The supervisor is responsible for the behavior of those in his department. He will be judged by the performance of others. If their behavior becomes unacceptable in terms of production or safety, the supervisor should analyze his behavior toward them.

Fortunately, much undesirable behavior is temporary. Behavior will vary from day to day, so the supervisor should vary his supervision so workers have some leeway within the limits of safe, efficient work.

Talking reduces strong emotions

Although no attempt can be made here to train supervisors to become personnel counselors, supervisors can help reduce the immediate result of strong emotion. You are in the best position to evaluate the emotional level of an individual worker and to do something about it.

Everyone is familiar with the emotional outbursts of people who have had a trying and upsetting experience. Such expression of emotion may take many forms from an outburst of invective, a torrent of tears, to physical assault on inanimate objects which cannot strike back. The end result is the same—relief from the intolerable tension, and emotional relaxation.

To reduce the strength of the emotional response, the supervisor can engage in what may be called permissive listening, that is, giving the employee an opportunity to talk about what is bothering him. This freedom to say what the individual pleases can reduce tension which might lead to distraction and accidents.

It is important that the supervisor not attempt to evaluate or judge the employee's remarks. The employee needs an opportunity to "blow off steam" without retaliation or evaluation.

The stresses which cause emotional upset can occur within the work environment or away from it, but each kind has its effect on the other. A simple statement such as "How are you today?" or "What's bothering you?" may well be sufficient to set off the safety valve of verbalization of pent-up emotion. After the discussions, the employee can often evaluate more clearly and realistically the things that upset him in the first place.

This relationship between the supervisor and the worker must be one of the supervisor's normal everyday functions. Diligent observation and effective and sensitive listening should be tools of every supervisor.

The "accident-prone" individual

Probably no phrase in safety causes so much disagreement as to what it exactly means than does "accident proneness." Most definitions hinge upon the idea that a person with certain personality traits is very likely to have accidents. When a person is said to be "accident-prone," it is generally meant that some psychological characteristics he has predisposes him toward having accidents.

Too often the term is loosely applied to anyone who has more accidents than others who do the same type work. A person could, however, have more than his share of accidents because he never was trained properly, or because he needs new glasses, or simply because he is working in cramped quarters or where he can be jostled.

It is true that a small group of people often account for more than their expected share of accidents during a given period, but over a long time period, the composition of the group changes—the accident repeaters of one time period do not usually show up during the next time period.

Unless it can be proved that accident-repeaters have certain characteristics, consistently linked to their accident repeating, it will be impossible to tell who is accident prone. It would be wonderful if such identification could be made, but at this time it is impossible.

More details are available in NSC's book *Supervisors Guide to Human Relations.*

The Alcohol and Drug Problem

Alcoholism is now classified as one of the most rampant illnesses plaguing modern man, and contributes at least 50 percent of all fatal motor vehicle accidents. Industry loses a sizable amount of profits each year due to alcoholism and drug abuse. Lewis F. Presnall, Director of Rehabilitation Services for the Kemper Insurance Group, commented that emotional disturbances cost American business approximately ten billion dollars per year. Of that amount, four billion dollars is due to alcoholism. Drug abuse accounts for a good portion of the remaining six billion dollars.

Interaction of alcohol and drugs

Studies indicate that at any one time from 10 to 20 percent of the population is taking prescription medication. Add to this the percentage of people using drugs illegally, and it becomes obvious that a substantial portion of our working population is exposed to the effects of drugs.

The *combination of alcohol and drugs* can produce a variety of effects that may severely impair a worker. Concentrations of alcohol

51

and drugs remain in the blood stream much longer than most users realize, and the effects of this combination may arise unexpectedly. Little scientific study has been conducted on the interaction of alcohol and drugs, but there is sufficient evidence to conclude that such a combination can lead to increased impairment of judgment and skill.

Alcohol and drugs on-the-job

In our society many of us have grown accustomed to the practice of having a cocktail or some other alcoholic beverage with lunch. Many individuals believe that a drink or two with lunch will not affect them. There may be no actual drinking on-the-job, but the effects of alcohol do not end with the meal.

Some individuals enjoy drinking, just as others enjoy eating, and no matter how often you speak about the impairing effects of alcohol, the person who enjoys drinking will continue to do so. If you are a supervisor, it is your duty to ensure that your workers reserve their drinking for off-duty hours. Also make sure that the effects of off-duty drinking do not accompany your employees on-the-job (in the form of a "hangover"). Remember that a hangover may also impair working and driving.

In addition to the multitude of prescription drugs being used both on and off the job, narcotics and dangerous drugs of all types are finding their way to work. The National Industrial Conference Board of New York conducted a nationwide survey of 222 firms and found that 53 percent reported cases of employee drug abuse.

One medical executive said that, unlike alcoholism, drug addiction is not amenable to ambulatory treatment. Whereas some companies report a 60 to 75 percent success rate with alcoholic employees, rehabilitation of an addict is fraught with a high rate of relapse. Drug addiction also spreads to other employees and leads to theft and to drug peddling. This is not true with alcoholism.

Countermeasures

Alcoholism and the sale and use of drugs by employees may not be problems in your company at this time, but will you be prepared to cope with these problems if and when they arise?

Alcohol. It is a small group of heavy or problem drinkers that is most tempted to drink during working hours. Even if you are fortunate enough to have no drinking on-the-job workers, remember that the problem drinker can still bring his hangover to work. Some studies indicate that employees who are alcoholics have more on-the-job and off-the-job accidents than non-drinkers.

Countermeasures for dealing with alcoholism begin with educating yourself and your workers. Information on alcoholism and the effects of alcohol is readily available. You may write to the following source for detailed information on alcoholism programs:

North American Association of Alcoholism Programs
1611 Devonshire Drive
Columbia, North Carolina 29204

You should try to obtain and read as much information as you can in order to be familiar with alcoholism and its effects. One good way to inform your people about alcoholism and the problems it can create is through the five-minute safety talk. You can also discuss drinking and driving during the next safety meeting.

The next step in an alcohol countermeasures program is the identification of problem drinkers under your supervision. Everyone is familiar with the drunk and can recognize the obvious signs of intoxication. The individual who has been drinking on-the-job will probably give himself away through his unsteady actions, or by the telltale odor on his breath. However, you should also watch for the apparently sober person with the "continuous hangover," which may indicate a serious drinking problem.

Having identified a problem drinker under your supervision, you should examine the ways in which you might help him overcome his problem. One way is to provide the individual with counseling by trained personnel who deal with alcoholism. Check your community organizations, churches, and local American Red Cross chapter for information on programs to assist alcoholics.

Above all, as a supervisor, you must show the alcoholic that you care and have a true desire to help him overcome his problem. Simply telling the alcoholic not to drink is like telling someone who enjoys eating, not to eat. An admonition not to drink is not the solution. Alcoholism is an illness and must be handled by persons trained to deal with it. The alcoholic needs help. It is up to you to convince him to accept treatment.

Drugs. The current increase in the use of drugs is not confined to the younger segment of the population. The individual who takes an overdose of aspirin or sedatives can be as much a problem as the heroin addict.

The Kemper Insurance Group has seriously considered the problem of employee drug abuse and has designed a policy to cope with it. Having had a high rate of success in dealing with alcoholism, Kemper Insurance modified and updated its safety policy to include what it terms "drug dependency." The Kemper Insurance drug abuse policy is as follows:

> In accordance with our general personnel policies, whose underlying concept is regard for the employee as an individual as well as a worker, we believe that:
>
> - Drug dependency is an illness and should be treated as such.
>
> - The majority of employees who develop a dependency on drugs can be helped to recover, and the company should offer appropriate assistance.
>
> - The decision to seek diagnosis and accept treatment for any suspected illness is the responsibility of the employee. However, continued refusal of an employee to seek treatment when it appears that substandard performance may be caused by an illness is not tolerated. Drug dependency should not be made an exception to this commonly accepted principle.
>
> - It is in the best interests of employees that, when drug dependency is present, it should be diagnosed as such and treated at the earliest possible stage.
>
> - *Confidential* handling of the diagnosis and treatment of drug dependency is essential.

The objective of this policy is to retain employees, who may have developed drug dependency, by helping them to arrest its further advance before it renders them unemployable.

The Kemper Insurance Group believes that identifying employees with drug problems is much easier if a company does not take the "discover and terminate" approach. Kemper Insurance maintains that all drug dependencies cause marked changes in work behavior patterns, personal relationships, and emotional moods. Its supervisors are instructed to watch for these changes in employees. When an employee is suspected of drug usage or dependency, his supervisor will privately counsel the employee and seek the assistance of the company staff or a community agency that deals with the problem and cure of drug dependency.

The Kemper Insurance Group has established a policy for dealing with illegal drug traffic:

"Any employees found to be involved in illegal drug traffic on company premises or during working hours will be terminated and reported to proper civil authorities."

With regard to employment of rehabilitated drug users, Kemper Insurance policy holds that:

"Persons who apply for employment and who can show satisfactory evidence of recovery from drug dependencies will not be discriminated against."

Your company must first establish a written policy when planning countermeasures to deal with drug abuse. This policy must be carefully prepared, written, and made known to all employees in the same manner as your company safety policy. The Kemper Insurance Group policy for dealing with drug dependency is excellent and could be adapted to fit your company.

After establishment of a company policy on drug abuse, you should locate any community or area organizations that offer counseling and assistance to persons with drug dependencies. You should familiarize yourself with the types of drugs that can lead to drug dependencies and the effects these drugs may produce. There is a vast quantity of material available on drugs and drug abuse. The following sources can provide you with this information:

National Coordinating Council on Drug Abuse Education and Information
P.O. Box 2000, Washington, D.C. 20013

The National Clearing House for Drug Abuse Information
5454 Wisconsin Avenue
Chevy Chase, Md. 20015

National Loss Control Service Corporation
Long Grove, Ill. 60049

As a supervisor, you should not strive to become a "drug detective," constantly looking for pills and drug substances or lurking in dark corners hoping to catch an addict giving himself a "fix." Instead, watch for changes in the work behavior patterns, personal relationships, and emotional moods of your workers. Any of these changes in an employee may indicate an alcohol or drug problem.

Summary

People have needs that require more than money to satisfy. These needs should be met insofar as possible if employees are expected to work safely and efficiently.

Everyone wants to feel his work has value and, when good, is appreciated sincerely. Everyone wants a sense of job security. Adequate wages and some permanence of tenure are basic to the feeling, but there are other factors too—factors over which the supervisor has a great deal of control.

A person should know what to expect from his supervisor, what his supervisor's reactions will be. He needs to know his supervisor does not play favorites and that he will be fair.

Finally, a person needs to be part of his work group. A supervisor can help this group be a cooperating unit in which each member has his place. New members will be more readily accepted because old members will not see them as potential threats to their security.

Above all, develop a sincere interest in and desire to help the employee with an alcohol or drug problem. Once you have developed these attitudes, you will want to help any employee solve his particular problem and again become a valuable asset to your organization.

All the answers cannot be given in one short chapter. In fact, no one could answer them all, especially when it comes to what a supervisor must say and do in every situation. But the supervisor who gains an understanding of those he supervises will become more skilled in handling all problems as they arise. His production will improve; his accidents will decrease.

CHAPTER 4

Maintaining Interest in Safety

Hazard control is management's responsibility. Top management formulates policy. The supervisor interprets it and makes it a reality to the workers. The supervisor is responsible for carrying out management policies in every way—and to the line employee, *he* is the company. This was discussed in Chapter 1.

The supervisor seeks to develop good attitudes in his people. He trains them in safe workmanship. He tries to convince them that management is sincerely interested in safety. He repeatedly emphasizes that he expects safe conduct.

These approaches are primarily problems in communications, problems in getting through to people. Communication means much more than tacking up posters, passing out rule books, and talking about safety. Success in safety depends on the attitude of both supervisor and employee, as well as top management.

How To Promote Safe Workmanship

The purpose of communication in safety work is to create and maintain safe attitudes through creating and maintaining interest in safety. The supervisor should use interest-getting techniques to encourage safe work practices and implement his safety program.

57

The supervisor is the key man in accident prevention—he personally inspires safe workmanship by his own attitude and example. Here the supervisor is reviewing safety performance.

Employee committees

One of the best ways to create and maintain this interest is to get employees involved in a well run safety committee.

One of the main purposes of an employee committee is to give employees an opportunity to take part in the operation of the safety program. The committee must have time to do its work and should receive acknowledgment for successful efforts. The committee's duties should be spelled out in writing.

Certain fundamental principles contribute to the success of an employee safety committee. Here are some of them.

1. A need must exist within the department, to which the employee safety committee can make a contribution.

2. Although the supervisor delegates various functions to the committee, he should retain the responsibility for departmental safety.

58

3. To carry out its assignments (inspecting, observing practices, investigating, and making recommendations), the committee must be given proper instructions, goals, and target dates.

4. The supervisor must communicate with members of his department. He should not use the committee as a buffer.

5. All committee recommendations should receive careful consideration. Worthwhile recommendations should be implemented. When the decision to adopt a recommendation is beyond the authority of the supervisor, it should be transmitted to a higher authority. If the final decision is negative, an explanation should be made to the committee.

6. Committee agenda should be limited to safety matters. The committee should not become involved in personnel or labor relations action or in any matter not related to accident prevention.

7. Committee membership should be rotated so that over a period of time, all will have an opportunity to serve.

8. Written minutes of all meetings should be kept.

9. Meetings should be held regularly as scheduled. Attendance must be required.

The conventional order of business for a safety committee meeting is the following:

1. *Record of attendance.*
2. *Consideration of unfinished business.*
3. *Review of recent accidents, including near misses.*
4. *Reports on special assignments.*
5. *Reports of inspections.*
6. *Progress report on safety program.*
7. *Special features such as a film, a talk or demonstration by a specialist, slides or a similar item.*
8. *Presentation of new business.*

A top management representative should occasionally be invited to a meeting. He can provide the "front line" point of view or explain new plans or policies. A summary of meetings should be sent to management.

Supervisors should make service on a safety committee interesting, productive, and rewarding.

Safety posters

Safety posters alert people to safe practices. A supervisor should use the company safety poster program and see to it that the few simple chores needed to maintain it properly are carried out.

Poster locations should be carefully selected. They should be in a prominent location that will not interfere with traffic, yet provide high visibility. They should be centered at eye level, about 63 in. from the floor. They should be in well lighted areas, with their own light if possible. (Never use a flashing light in a production area.) A good size for the poster board is 22 in. wide by 30 in. long, or it may be only big enough to hold one poster. Standard National Safety Council posters are available in two sizes: "A" size—8½ by 11 in. and "B" size—17 by 23 in. Poster boards or frames should be attractively painted and may be glass covered. One board is usually desirable in the work place. In wash rooms or locker rooms or where people bring their lunch to eat, several panels may be used effectively.

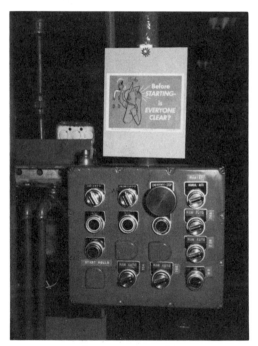

"POP" poster in place.

Posting material should be displayed alone and kept free of clutter or other notices.

Frequent change of posters and display material is preferred. Add a few new posters; circulate other posters. Selection and rotation of posters can be handled by the safety director. If the supervisor selects posters, he should choose those that keep employee interest

The Insiders—a safety promotion in which certain employees were authorized to deliberately, but carefully, perform unsafe acts—such as riding in the forks of lift trucks, to blocking walkways with paint buckets. Those who cautioned the "insiders" against these acts were given cash awards on the spot. No one knew who or where the "insiders" might be, so everyone was suspect. Result—no lost-time accidents among 1600 employees during the period of the promotion.

alive. Posters may provide notice of forthcoming holidays, and other special subjects.

Intensive use of posters covering certain bad practices may be effective for punching hard at certain outstanding hazards in the department. If poor housekeeping is a persistent problem, for instance, the poster program could be used to reinforce the planned cleanup activities. Homemade posters can be effective.

Special-purpose reminders

The National Safety Council's "POP" (point of problem) posters and safety stickers will highlight particular hazards in an area. POP posters are 4¼ by 5½ in., and stickers are 2½ by 3¾ in. Both are self-sticking.

Occasionally there is material that must be on permanent display. A poster describing use of fire equipment or respirator, or a warning for radioactive material, comes under this category. Such posters

The dummy tells the story. Sign on dummy's chest says, "Let's make this year ACCIDENT-FREE, or your cuts, bruises, and bandages will show on me!"

should be mounted under glass or on heavy board to make them permanent. Such posters should be reviewed regularly to make sure they are not out of date.

Poster contests can be used to create interest in safety.

Off-beat ideas

A change-of-pace display might be tried occasionally.

To dramatize the effectiveness of safety hats, for example, an impact-testing device might be used, or the hat could be mounted on a dummy and a heavy blow struck against the hat at repeated intervals. Displays of safety glasses, safety shoes, or safety hats that have prevented injury to their wearers are very effective.

The holding power of a safety belt is demonstrated in this exhibit.

In recognition of receiving NSC's Award of Merit, a "mobile celebration" toured the plant. Coffee and doughnuts were served to all.

63

A remodeled 1925 caboose brings Santa Claus to children of DT&I railroad's employees and the general public. Children talk with Santa and receive a small gift. Parents are given a Christmas safety booklet.

To show how much accidents cost, one firm ran a guessing contest as to how much it cost a man to be off the job with an injury. The average wage and average time away from work were used. When the figure was posted and the winner was named, the actual amount was shown.

Sometimes a simple interest-arouser is best. One supervisor promised to pay for a round of coffee, tea, or milk to everybody in his department on the day that department could ring up a half-million man-hours without a disabling injury. He paid off, and was glad to do it.

Recognition organizations

There are many organizations throughout the United States and Canada that have awards for people who have eliminated a serious injury or at least minimized an injury through the use of certain articles of personal protective equipment. The award provides an excellent opportunity for publicity.

Some of these recognitions are:

Wise Owl Club. Founded in 1947, this is the oldest of all such "clubs." The award is made to one who has saved his eyesight because he was wearing eye protection. Address inquiries to Director of Industrial Service, National Society for the Prevention of Blindness, Inc., 79 Madison Ave., New York, N.Y. 10016.

Turtle Club. This award is made to one whose life was saved or serious injury prevented because he was wearing a safety hat. Address inquiries to Turtle Club, 2680 Bridgeway, Sausalito, Calif. 94965.

Scarab Club, Ole Safety Sam Club, and *Golden Shoe Club* are all awards made to employees who avoided serious injury because they were wearing safety shoes. Address inquiries to Scarab Club, Emmaus, Pa. 18049; Old Safety Sam Club, 9th and Greenleaf Streets, Allentown, Pa. 18105; or Golden Shoe Club, 1509 Washington Ave., St. Louis, Mo. 63166.

Half-Way to Hell Club award is made to employees who have been saved from death or serious injury by falling into safety nets installed at a job site. Address inquiries to Half-Way to Hell Club, 533 Second St., San Francisco, Calif. 94107.

Kangaroo Club members are individuals who have averted serious injury or death by wearing vehicular safety belts, when driving or as a passenger, either on or off the job. Address inquiries to Kangaroo Club International, P.O. Box 950, Coatesville, Pa. 19320.

Safety contests

Safety contests are popular. About five million American workers participate each year in the industrial safety contests sponsored by National Safety Council members. More than nine billion man-hours are totaled each year. Hundreds of awards are presented.

This all pays off, too. Contests can be one of the most effective means ever devised for creating interest in accident prevention. The supervisor who actively supports his company's contest, or who starts a department contest, is making use of one of the best safety motivations known.

Competition may be between departments, plants, divisions, or

When this mill reaches certain goals for accident-free days, scholarships for employees' children are raffled. Money is deposited in company credit union for withdrawal when child reaches college age.

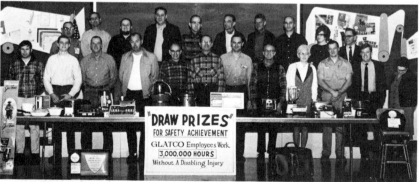

Three-million man-hours without a lost-time accident. Company recognized the event by giving all employees a home tool kit. Others (pictured above) drew lucky numbers for larger prizes.

companies within an industry. In addition to the nation-wide contest sponsored by the National Safety Council there are contests sponsored by trade associations and local safety councils.

Safety contests, it must be emphasized, are not a substitute for a safety program. Safety contests have no other purpose than creating interest in an organized safety program. If they fail to do this, they are worthless.

Details. Contests are based, generally, on accident experience and are operated over a period of six months or a year. Winners are determined by relative standings, by performance improvement, or by other factors agreed upon in advance. Many effective local contests are in the nature of special campaigns, launched with advance publicity, and run for a specified time.

Inter-department contests challenge the supervisor the most. (National inter-company contests usually are directed by safety department personnel.) Since workers are more likely to have a personal interest in the standing of their own department, this type of competition has been most successful.

When inter-departmental contests are conducted between dissimilar departments in one plant, difficulties created by differences in size, type of operation, and exposure to hazard must be overcome.

• One solution is to judge the percentage reduction in each department's frequency rate, from a previous base period (usually the previous year).

• Another method is to handicap high-hazard departments on the basis of the compensation insurance rates for the different operations. If, for example, the rate per $100 of payroll is $3.00 for Department A, $2.00 for Department B, and $1.50 for Department C, factors for the units are 3, 2, and 1.5 respectively. To adjust the frequency rate of each department, divide the frequency rate by the handicap.

When two or more departments come up with perfect records, the department with the most man-hours worked since the last chargeable injury can be named the winner.

For contest purposes, the basis for calculating frequency rate and defining chargeable injuries is the American National Standard Z16.1, *Method of Recording and Measuring Work Injury Experience.* Also see the explanation of incidence rates on page 37.

When departments vary greatly in number of persons employed and in the nature of work, teams of from 20 to 50 people each can participate in intergroup competition. To equalize the variables, each team is made up of a proportionate number of people from high-, medium-, and low-hazard occupations. Interest is increased when teams are named after prominent football, baseball, or other outstanding sports teams, and the entire competition named after a league or other sports organization. Teams names can be chosen by the group or drawn from a hat. Colored buttons can designate team members.

Less-formal contests among departments are also feasible. Contests can be held for good housekeeping, greatest improvement in housekeeping, or wearing personal protective equipment. Housekeeping contest winners are determined through unbiased inspections made periodically. Demerits are given for conditions agreed in advance to be undesirable.

Sometimes elaborate point systems are worked out so that, in addition to helping win the department trophy, employees also accumulate "points" for the department performance. When scores are totaled, employees may select merchandise from a catalog, on the basis of points accumulated. Sometimes trading stamps are used for this purpose.

Increasing use is being made of personal no-accident awards, such as pencils, engraved buttons, billfolds, and similar articles, to acknowledge employee cooperation in working for the contest. Some companies hold raffles of merchandise or cash for employees who have not suffered disabling injuries.

Some companies also recognize employees who have long-term safety records. They are presented with a watch, and sometimes even with a testimonial dinner.

Should all this activity seem pointless to a supervisor looking for means of motivating interest in safety, he should consider that in cold print it may seem so, but in actual practice it is highly successful if handled right. Since the purpose of contests is to stimulate interest, there is virtually no limit to the number of ways in which contests can be improved by a little showmanship. (See the previous section on "Off-beat ideas.")

One of the ways that a contest is handled right is to make sure that recognition is positive, rather than negative. The supervisor should make sure his people are rewarded for good safety perform-

Suggestions are encouraged by placing posters and suggestion form racks in employee lounges and cafeterias. Company distributes an information booklet that gives complete details of the suggestion plan. Company publicizes award-winners—maximum award at Parke, Davis & Co. is $15,000.

ance, not ridiculed or publicized for bad safety performance.

Suggestion systems

Many people, from time to time, think up ideas that would help the department or the company improve work methods or reduce work hazards, or both. All too often, such ideas are lost because there is no effective way to bring them to light.

A well organized suggestion system successfully encourages em-

Red Cross First Aid training is given to pipeliner, communication, supervisory, office, and management people at Laurel Pipe Line Company, Pennsylvania.

ployees to contribute their ideas and, in addition, stimulates their thinking about solving problems of production and safety. Solutions of production problems are rewarded by giving the employee a percentage of the tangible savings. Solutions of safety problems are rewarded according to the importance of the idea.

A committee, under the leadership of an executive, receives and evaluates the ideas. Some companies have suggestion departments or use the services of professional suggestion administrators and organizations.

Usually, special forms are provided for submitting suggestions, and special boxes provided for their collection. Suggestions should be gathered frequently and acknowledged promptly. Results of decisions should be given right away, with a brief explanation if the suggestion is not accepted.

The supervisor is usually not eligible to win awards for suggestions. He should let his people know that he will help them work out their ideas. He should make it clear that no suggestion will be considered as criticism, nor will it result in discrimination against the person making the suggestion.

The successful suggestion system must have a clearly stated and easily understood operating plan that is scrupulously observed. Employees must know the plan is fair, impartial, and potentially profitable to themselves. The supervisor can do a great deal to encourage that faith by the way he administers the suggestion system in his own department.

First aid courses

Companies that have conducted a good first aid course have found safety to be a conspicuous by-product. Some companies include first aid in their employment training program. The value of first aid training is most obvious in operations far removed from professional medical help.

Such training should be standard Red Cross courses or the Bureau of Mines courses. The need for personnel trained in first aid is spelled out in OSHA requirements if there is no infirmary, clinic, hospital, or physician in near proximity or reasonably accessible.

Safety meetings

Safety committee meetings were discussed earlier in this chapter.

Department meetings can feature a film or a chart talk on a subject related to the work of the group or to off-the-job safety. Models or exhibits, safety equipment, or actual equipment can be used.

Visual aids with pictures and lettering that can be seen by a group of 15 to 20 people are easily found or made for a wide variety of subjects. Here is how.

A supervisor can produce his own flip chart at little trouble or expense. All that is needed is a dozen or so sheets of white paper (or a large pad), ranging from 18 by 24 in. to 24 by 36 in., clamped at the top to a backing board, some markers (felt-tip, crayon, or colored chalk), and an outline of points to be discussed. Details are given in NSC Industrial Data Sheet No. 564, *Nonprojected Visual Aids.*

A set of 35mm color slides (or film strips) is a superior training aid. They can be purchased or made. See details in NSC Data Sheet No. 574, *Projected Still Pictures.* The Council offers sets of professionally produced 35mm slides on a wide variety of industrial safety subjects. Each set has a reading script. Some have audio-cassettes.

Motion pictures are excellent for larger groups, if the projection equipment and viewing area are available.

All pre-prepared visual aids are just that—aids. They are powerful training and motivating devices but are not intended to put across

Safety education may be taken to work crews in the field. This "Ciné-O-Van" uses rear-screen-projected slides and movies to bring safety instruction to field men in this sugar company's operation. Railroads use similar units to reach maintenance-of-way gangs. Construction and highway maintenance operations also find mobile visual aid units to be valuable.

a message all by themselves. They lose much of their value if they are not used to reinforce a discussion, or to serve as the basis for a discussion. They should be selected because they fit; not used only because they are available.

The larger meeting, involving more than one department, is valuable when company policies need explaining or when general accident causes are discussed. Some meetings may be purely inspirational, to create an awareness of hazards and the desire to prevent accidents. Frequently they are held for presentation of safety awards.

The "production huddle" is another type of meeting—an instruction session about a specific job. Although not restricted to safety, it deals with the problem of getting the work out safely. The supervisor who holds one of these sessions and who gives proper emphasis to safety may well feel that he has had the best kind of safety meeting. Safety has now been "built in," which is what it should be.

Public utility crews use this type meeting. They call it a "tailboard conference." Before tackling a job, the crew gathers around its truck and discusses the work. They layout the tools and materials they will need and agree upon which part of the job each man is to handle. Such sessions are indispensable, especially in highly hazardous situations. Train crews and switching crews can also use this type of get-together.

How to hold a meeting. A safety meeting, like any other, will most likely succeed if the person running it is determined that the audience will leave with more useful information than it had when it came in, and will leave with an improved attitude toward the subject at hand. This can happen if the supervisor knows a few facts about human beings.

1. People do not like to go to meetings unless they anticipate being interested or entertained.
2. A person who is personally and passionately dedicated to safety will undergo great boredom in the interest of safety—but, unfortunately, most people are not that dedicated.
3. People can absorb only a few ideas at a time. The meeting should aim to drive home one or two, certainly no more than three, conclusions.
4. People become irritated and unreceptive when they are uncomfortable. They become uncomfortable if they have to stand too long, if they are too warm or too cold, if they have to look into strong light, if they must listen over background noises, or if they otherwise feel ill at ease.
5. No matter how good a meeting may be, people get a bad impression if the program has been mismanaged. To avoid this, good planning is necessary. The film should be previewed to make sure everything runs properly and the people running the meeting can make intelligent comments about it. Other visuals should be practiced. Talks should be rehearsed and timed. Exhibits should be checked to make sure they work properly.

Unless there is some good inducement, an absorbing film, or some other "good hook" on which to hang its interest, a safety session should not run longer than 30 minutes—keep it short and simple. Only the most exceptionally appealing material justifies an hour's get-together, even if the time is available. These estimates, of course, do not apply to instructional courses, but, rather, group meetings.

The first question a supervisor may ask when he is about to hold a safety meeting with his people is, "On what subject?" The safety director or safety department should be able to provide a multitude of ideas, either based on company experience or secured from the National Safety Council (see "Useful References").

The supervisor who cannot himself answer the question, "What subject?" is the supervisor who has absolutely no problems of safety, either on the job or off it. If he starts with his own problems and then looks to those of his people, he will not need to look long for a subject.

Off-the-Job Accident Problems

The main goal of safety is to prevent accidents and injuries—whether on or off the job. The off-the-job problem may be simple, but it is big. Refer to the table. In a recent year, accidents away from work accounted for over 70 percent of all deaths and over 55 percent of all injuries to workers. As company safety programs become more effective, at-work deaths and injuries account for a smaller share of the total injury experience.

These figures are for all industries. In some industries, disabling injuries occur 10 to 20 times as often off the job as on the job. Even while many companies are building an outstanding work safety record, their employees go right on suffering disabling injuries and death while off the job.

An employee is safer *on* the job because these four basic principles of accident prevention are continuously practiced:

1. Every effort is made to match the person to the job.
2. Employees are trained and motivated to do their jobs the safe way.
3. Tools, protective equipment, machines, and the work area are maintained in first-class condition.
4. Materials are handled according to safe procedures, industrial hygiene is practiced, and machines are safeguarded.

When the employee leaves his job to go home, he leaves behind this carefully constructed network of safety. He is now part of the "general public" and as such faces the hazards common to the general public—on streets and highways, at home, and at play. He is now his own boss. His injury experience is now a measure of his ability to supervise himself and his family intelligently.

Companies must educate, motivate, and continually remind em-

TABLE II		
ACCIDENT TOLL IN A TYPICAL YEAR		
Gainfully employed persons only		
Place	*Deaths*	*Injuries*
All Accidents	55,700	5,500,000
At Work	14,200	2,300,000
Away From Work	41,500	3,200,000
Motor vehicle	25,100	900,000
Public nonmotor vehicle	8,600	1,100,000
Home	7,800	1,200,000
Source: NSC *Accident Facts*, 1972 ed.		

ployees of the importance of following safe practices off the job, as on the job. This is a real job for the supervisor. It is another place where he must set the example.

Interest in OTJ safety makes a sound investment. Firms that attack the total problem of accident prevention attest that OTJ safety pays off handsomely by reducing the incidence of injuries both on and off the job.

Cost of off-the-job accidents

The cost of off-the-job accidents each year is about twice the $9 billion annual cost of on-job accidents (figures include property damage as well as personal losses, wage losses, claims, medical and hospital costs, and administrative expenses). It affects production costs and schedules, because skilled manpower has been lost.

If replacement workers must be hired, personnel department costs, wages spent for training, and slower production must be figured. The cost of tools damaged and materials spoiled by new and inept workers must not be overlooked. Finally, wages paid to noninjured persons for

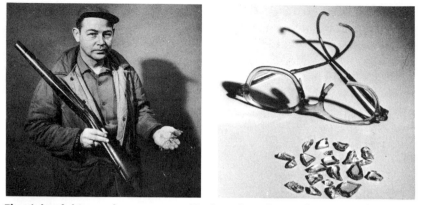

The sight of this employee was saved by his safety glasses when his rifle bolt back-fired on a hunting trip. Many workers, including office employees, take advantage of their company's connections with safety glass distributors and get the extra protection of this equipment off the job as well as on the job.

nonproductive time (such as court appearances, hospital visits, cleaning up) as the direct result of an off-the-job (OTJ) accident must be added.

Personal cost to the injured man or woman is often beyond measure. Suffering, loss of earning power, and an upset home environment —all must be reckoned with.

Mental anguish may be as much to blame as is physical impairment for a man's reduced efficiency when he does return to work. Worry and anxiety can lead rapidly to an alarming neglect of safety. The result may well be an increase of on-the-job accidents (also called "on-job" accidents).

Accident-free employees work more days, and are more efficient when at work. They spoil less materials, do less damage to tools and machines, and have higher rates of production.

Humanitarian reasons for OTJ safety activities

The humanitarian reason is, of course, a primary one. No one wants to see his friends and associates suffer. OTJ safety activities help build good employee and good public relations. They are substantial evidence that a company believes in the dignity and worth of each employee.

A comprehensive OTJ safety program deserves the vigorous support of every company—large or small, no matter what product or service it produces, no matter how "safe" the company might be.

Measuring accident experience

The off-the-job frequency formula is the standard method of measuring and comparing OTJ accident experience.

$$\frac{\text{Number of OTJ injuries} \times 1,000,000}{312 \times \text{Number of employees}} = \text{OTJ frequency rate}$$

The formula is based on 312 exposure hours per employee per month. Here is why. An employee normally works eight hours a day, five days a week. The eight hours per day for sleeping are excluded. This gives a person eight hours of exposure per weekday and 16 hours on Saturday and Sunday. This adds to 72 exposure hours per week, multiplied by 4⅓ weeks per month, for a total of 312 exposure hours per month.

No adjustment need be made for overtime since it will be offset by holidays, vacations, and incidental absences.

For example, a plant with 10,000 employees has five OTJ injuries during one month. Each injury causes only one day's absence from work. The OTJ frequency rate for that month will be:

$$\frac{5 \times 1,000,000}{312 \times 10,000} = 1.60$$

To be reportable, an OTJ injury must result in a person's losing at least one full day, be it a working day or a holiday, weekend day, or a vacation day. No medical opinion is needed. Fatal or permanent injuries are also included.

No scheduled time charges are made for permanent partial disabilities or fatalities. Only the number of lost-time injuries and the number of fatalities are reported.

OTJ injuries are classified under three types: transportation, home, and public. As described in NSC Form 1100–2, "Quarterly Summary, Off-the-Job Accidents" (available from OTJ Safety Director, National Safety Council), these classifications cover the following types of injuries.

Transportation—Injuries caused by or resulting from accidents at places other than home premises, involving a moving automobile,

"On-the-job safety begins here"—at entrance gate to railroad shop.

truck, bicycle, bus, streetcar, motorcycle, railroad, boat, airplane, and pedestrian when a moving vehicle is involved. Classification also includes injuries occurring during boarding, alighting, or moving within vehicles.

Home—Injuries incurred in the home or home yard area, caused by vehicles, firearms, machinery, tools, fire, explosion, exposure to heat or cold, electricity, toxic material, fall, slip, improper lifting, hot object or material, sharp object, striking an object, over-exertion, or animal, insect, or other causes.

Public—Injuries other than "transportation" or "home," caused by firearms, fire, explosion, fall, slip, striking objects (including stationary vehicles), animal or insect cause, fight, assault, exposure to heat or cold, sports, toxic material (such as poison ivy), electricity, or other cause.

By having subdivisions in these three classifications, companies can evaluate the effectiveness of program areas and pinpoint where and when the greatest efforts are needed.

Other side of entrance sign tells that off-the-job safety starts at same spot.

Integrating OTJ with on-the-job safety programs

A company safety program cannot be completely effective unless it deals with the "whole" person—not just the person as a machine operator or a truck driver. A person is not likely to be as safe at work when he disregards safety as soon as he walks out the door or the plant gate. Conversely, if a person is careful at home and on the highway, he is likely to be receptive to safety education as an employee.

This is why more and more employers are directing their attention to OTJ safety education, and are integrating it completely with on-the-job safety education.

Here is how.

- Many organizations provide reading racks stocked with helpful literature for employees to take home.

- Some send material directly to an employee's home. Traffic, vacation, and home safety leaflets and other material are available. Over a million copies of National Safety Council's *Family*

79

"Defensive Driving" Courses, such as those developed by NSC, are intended to teach better driving to millions of automobile drivers. Many firms sponsor these courses to improve the skills of both on-the-job and off-the-job driving.

Safety are subscribed to by companies for their employees.

- Companies sponsor safety-oriented activities: the National Safety Council's "Defensive Driving Program," Red Cross first aid courses, learn-to-swim campaigns, and even "family night" safety rallies.

Support for community and family safety programs reinforces a company's safety efforts. Company officials, supervisors, and employees should participate.

The supervisor's role in OTJ safety

OTJ safety activities affect the general well-being of employees, the operation of the company safety program, and the efficiency of the supervisor's own department. The methods used to promote these activities are the same as those used to promote on-the-job safety.

- The supervisor must believe in and practice safety all the time, wherever he may be. He must recognize its importance to the well-being of his own family. He must set an example.

- The supervisor should determine if it were an injury that kept a person from work. He should encourage reporting of all accidents. Even reporting of near-misses can supply topics for safety discussions.

- Part of department and committee meetings can be devoted to discussing OTJ safety. Supervisors can talk individually to people about it.

- Bulletin boards can include vacation or holiday safety hints— and clippings that may be of interest to employees.

- The supervisor can support public and home safety activities. Local safety councils, churches, fraternal organizations, and youth groups need safety-minded volunteers.

CHAPTER 5

Instructing for Safety

People do unsafe things because:

- They are unaware that what they are doing is wrong;
- They misunderstand instructions that are given;
- They do not consider the instructions to be important;
- They are not given specific instructions;
- They find it awkward to follow the instructions; or
- They deliberately disregard instructions.

Often they are unaware that what they are doing is wrong because they lack the knowledge or skill to do their job properly and safely. This is one of the most frequent personal causes of accidents, and may be due to one of these reasons:

- The person may never have learned to do his job the right way;
- He may have learned it once, but not well enough for correct, safe work habits to stick with him; or
- He may have learned how to do the job safely under normal conditions but has never fully realized the danger of certain unsafe acts, possibly involving some unusual job conditions.

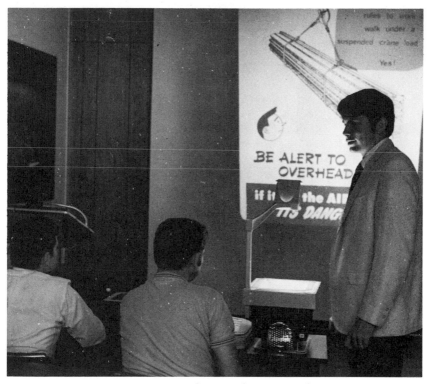

New employee orientation uses visuals to emphasize the safe way.

Good instruction is important to safety. This chapter discusses: how the supervisor can best give instruction, the need and importance of job safety analysis (JSA), methods of job safety analysis, job instruction training, and how all of this relates to accident prevention.

Giving Instruction

One job a supervisor really needs to do well is giving instruction. Because he usually does not turn out production himself, he works through people. He directs them and they follow his instructions. If these instructions are good, the supervisor can expect the work to be done much as he wants it. If they are bad, incomplete, or unclear, the supervisor can expect poor work, mistakes, and accidents.

Highlights of a company's safety program are presented to a research management team. New employees aren't the only ones who need safety instruction.

Importance of job instruction

Each person starting a job should be instructed to do it properly. Even if he has performed similar work, he needs to know what is expected of him now, because each supervisor has his own ideas of how a job should be done. A safety-conscious supervisor wants to make sure the new person is going to work safely, and this might not have been emphasized where the person worked before.

Good instruction contributes to safety, even if the instruction is not specifically on safe practices. Good instruction makes a man a better worker, more satisfied with his job and his company, and more willing to go along with programs which the company and his supervisor advocate. The man who knows what is going on listens when his supervisor talks about safety. He makes suggestions. If he sees a hazard, he tells his supervisor about it. When men feel this way, there are fewer accidents.

Who instructs? Sometimes a supervisor gives job instruction himself, but often he has someone else do the teaching. He may be too busy with other work to be able to instruct every learner. It may be true that somebody else, an assistant foreman or an experienced worker for example, may have more first-hand knowledge of the job than does the supervisor.

Regardless of who does the breaking-in job, it is still the supervisor's responsibility to know what is being taught and whether or not the learner is really mastering it. He should occasionally observe the instructor at work, and check each learner's progress. This keeps him informed, and shows his interest as well. The learner must feel that, although his supervisor does not give the instruction, he is responsible for it—that whatever is taught has his approval.

When a supervisor chooses experienced employees to act as instructors, he should coach them in how to go about the teaching job. He should observe the results of their work, as borne out by the quality of the work produced by the people they instruct.

Some experienced men show special liking and aptitude for instructing. They can be given extra coaching in training and leadership; they could be likely prospects for supervisory jobs.

Starting the new man

At the time he is employed, even before he starts his job, the new man should be told something about company and department safety policies. This can be brief, but must be the good start that will make later instruction and supervision easier.

A new man wants to get started on his job. He can be worn down by too much talk, much of which he cannot understand until he has seen what the job is like. No matter how brief this initial talk, some things must be said. The supervisor should think about them carefully.

- Management and employees are determined to prevent accidents, because accidents are bad for all concerned.

- Most employees here have never suffered a disabling injury at work, except in rare instances.

- In spite of everything the company has done, people sometimes get hurt. To avoid getting hurt, each person must look out for his own safety.

- Job instruction will include safety instruction. (No one is expected to do a job he does not understand. People are urged to ask questions about any part of their work they do not understand, and supervisors will answer questions willingly.)

- Instruction is not fault-finding; correction is not reprimanding.
- It is positively forbidden for any person to try to operate any mechanical equipment without instruction and specific authorization from his supervisor.
- Persons are urged to report anything about the work that seems unsafe.
- All injuries must be reported to the supervisor.
- Methods of procuring protective equipment, and rules and practices governing its use, will be explained.

Often the safety director or personnel director talks to new employees about safety before they are directed to their departments. This is good practice. It shows management interest in safety and strongly supports the safety instruction given later by the supervisor. This practice is no substitute for instruction by the supervisor.

The supervisor should, of course, know what has been said in the employment office, but even if all the items listed before were talked about, he should cover them again. An employee must never be allowed to think that safety is something he hears about in the office but seldom on the job.

On-the-Job Training

On-the-job training (OJT) is widely used because the trainee can be producing while he is being trained. Whether the supervisor does the instructing himself or has someone else do the teaching, the training should be carefully planned and organized.

In too many cases, on-the-job training is a "hit or miss" procedure where the trainee is told to follow another worker around and learn his job. In situations of this type, the worker may be too busy to do any training, or may not know good training procedures. He may even be reluctant to train another to do his work.

Remember, it is the supervisor's responsibility to make sure the person doing the training can, in fact, train. On-the-job training in-

Over-the-shoulder coaching. The trainee is expected to develop and apply skills in the work situation under the guidance of a qualified person.

cludes many techniques and approaches and there is no one method that will fit all situations. Job safety analysis (JSA) is one method, job instruction training (JIT) is another, and over-the-shoulder coaching is still another widely used method. These methods may be used separately or in combination depending upon the complexity of the job and the time element.

Over-the-shoulder coaching

Over-the-shoulder coaching is perhaps the most flexible and direct of the three training methods. In coaching the trainee is expected to develop and apply his skills in typical work situations under the guidance of a qualified person. The person to whom the trainee has been assigned should be one who knows the job thoroughly, is a safe operator, and has the patience, time, and desire to help others.

Date March 1
Supervison L.O.
Department office

Training chart — Tasks (CASHIER / CLERICAL) by Employee

Employee	KNOW CUSTOMERS	CUSTOMER RELATIONS	ACCURACY IN CASH	MAKE UP BANK DEPOSIT	TAKE DICTATION	STENO	COMP OPERATOR	FILE	MAIL	TELEGRAMS	PLANT ORDERS TELEPHONE BOARD	Notes
1. Mrs A.M.	X	X	X	X			⊗		?			FREQUENTLY ABSENT. MISS B.C. HAS TO TAKE OVER WORK
2. Miss B.C.	X	X	X	X	X	X	⊗					
3. Miss W.E.	X	X	X	X	NQ	X	X				X	TRAIN HER IN CASH AND DISBURSEMENT JOURNALS
4. John D.			X	X	X	X	X	X	X	X		CAUTION ON MORE CARE IN CHECK WRITING ON PLANT PAYROLL
5. Mrs W.C.	⊗	⊗	⊗	⊗				X	X			TRAIN HER TO HELP ON FREIGHT BILLS
6. Miss E.M.					⊗	⊗	⊗	⊗	⊗		⊗	START HER ON CHECKING BANK STATEMENT. ASK W.E. TO HELP
7. Miss J.O.								⊗			2/5	GET HER STARTED TO HELP ON TELEPHONE BOARD
8. Miss M.O.			?		X		⊗	⊗	X			TRAIN ON STENCILS
9. Miss K.B.							X	⊗	X			GOOD PERSON'TY — HOW ABOUT WORKING TOWARD CASHIER AIDE
10. Miss L.B.							⊗	B/W	B/W			
11. Bill K.				⊗								WORK WITH HIM ON MAIL — RE-ARRANGE-FIND BETTER WAY

CODE
NQ – Not qualified
X – can do
⊗ – Better way?
? – Check it can do
⊗ – Regular Job OK
2/5 – Target date

Training chart keeps track of who needs (and who has) what training.

The advantages of training of this type are:

1. The worker is more likely to be highly motivated because the guidance is personal.
2. The instructor can identify specific performance deficiencies and take immediate and proper corrective action.
3. Results of the training are readily apparent since real equipment is being used and finished work can be judged by existing standards.
4. The training is practical and realistic and can be applied at the proper time.

Timing is important; not only do the trainees like to get help when needed, but also the instructor can judge the trainee's progress continually so he can present the next unit or phase of instruction when the trainee is ready.

To help keep track of the progress of each individual, a training chart (shown) is valuable. On the top of the chart are listed the various tasks required of a person in a particular job classification. By observation and direct discussions, the instructor can determine whether the employee is qualified to perform the tasks necessary to fill the job. The instructor determines the degree of skill and knowledge of the trainee and estimates his training needs. He makes notes and comments on the chart. Such a chart prevents neglecting training that is important; it also prevents unnecessary training.

Job Safety Analysis

A job is a sequence of separate steps or activities that, when put together, accomplish a work goal. Identifying hazards and stressing safe procedures are part of job safety analysis and should be one of the first steps taken where there is a possibility of injury to the worker.

Here are the steps to follow when analyzing a job.

Select the job

Jobs suitable for a job safety analysis (JSA) are those assignments that a line supervisor may make. Operating a machine, tapping a furnace, piling warehouse goods are all subjects for JSA. They are

JOB:			DATE:
TITLE OF MAN WHO DOES JOB:		FOREMAN/SUPR:	ANALYSIS BY:

JOB SAFETY ANALYSIS TRAINING GUIDE

		REVIEWED BY:
DEPARTMENT:	SECTION:	APPROVED BY:

REQUIRED AND/OR RECOMMENDED PERSONAL PROTECTIVE EQUIPMENT:

SEQUENCE OF BASIC JOB STEPS	POTENTIAL ACCIDENTS OR HAZARDS	RECOMMENDED SAFE JOB PROCEDURE
Break the job down into its basic steps, e.g., what is done first, what is done next, and so on. You can do this by 1) observing the job, 2) discussing it with the operator, 3) drawing on your knowledge of the job, or 4) a combination of the three. Record the job steps in their normal order of occurrence. Describe what is done, not the details of how it is done. Usually three or four words are sufficient to describe each basic job step. For example, the first basic job step in using a pressurized water fire extinguisher would be: 1) Remove the extinguisher from the wall bracket.	For each job step, ask yourself what accidents could happen to the man doing the job step. You can get the answers by 1) observing the job, 2) discussing it with the operator, 3) recalling past accidents, or 4) a combination of the three. Ask yourself: can he be struck by or contacted by anything; can he strike against or come in contact with anything; can he be caught in, on, or between anything; can he fall; can he overexert; is he exposed to anything injurious such as gas, radiation, welding rays, etc.? for example, acid burns, fumes.	For each potential accident or hazard, ask yourself how should the man do the job step to avoid the potential accident, or what should he do or not do to avoid the accident. You can get your answers by 1) observing the job for leads, 2) discussing precautions with experienced job operators, 3) drawing on your experience, or 4) a combination of the three. Be sure to describe specifically the precautions a man must take. Don't leave out important details. Number each separate recommended precaution with the same number you gave the potential accident (see center column) that the precaution seeks to avoid. Use simple do or don't statements to explain recommended precautions as if you were talking to the man. For example: "Lift with your legs, not your back." Avoid such generalities as "Be careful," "Be alert," "Take caution," etc.

neither too broad nor too narrow.

Jobs should not be selected at random—those with the worst accident experience should be analyzed first if JSA is to yield the quickest possible return. In fact, some companies make this the focal point of their accident prevention program.

Break the job down

Before the search for hazards begins, a job should be broken down into a sequence of steps, each describing what is being done. Avoid the two common errors: (a) making the breakdown so detailed that an unnecessarily large number of steps results, or (b) making the job breakdown so general, that basic steps are not recorded.

The technique of making a job safety analysis involves these steps:

1. Selecting the right man to observe.
2. Briefing him on the purpose.
3. Observing him perform the job, and trying to break it into basic steps.
4. Recording each step in the breakdown.
5. Checking the breakdown with the man who was observed.

Select the man who is experienced, capable, cooperative, and willing to share ideas. This fellow will be easy to work with.

If the man has never helped on a job safety analysis, explain the purpose—to make a job safe by identifying hazards and eliminating or controlling them—and show him a completed JSA. Reassure the man that he is not being watched to see if he works safely or not, but that the job is being studied, not him. Reassure him that he was selected because of his experience and capability.

To determine the basic job steps, ask "What step starts the job?" Then, "What is the next basic step?" and so on.

To record the breakdown, number the job steps consecutively as illustrated in the first column of the JSA training guide, illustrated. Each step tells what is done, not how.

The wording for each step should begin with an "action" word, like "remove," "open," or "weld." The action is completed by naming the item to which the action (expressed by the verb) applied, for example, "remove extinguisher," "carry to fire."

91

In checking the breakdown with the man observed, the supervisor should obtain his agreement of what is done and the order of the steps. The supervisor should thank the man for his cooperation.

Identify hazards and potential accidents

Before filling in the next two columns of the JSA—Potential Accidents or Hazards and Recommended Safe Job Procedure—begin the search for hazards. The purpose is to identify all hazards—both those produced by the environment and those connected with the job procedure. Each step, and thus the entire job, must be made safer and more efficient. To do this, ask yourself these questions about each step:

1. Is there a danger of striking against, being struck by, or otherwise making injurious contact with an object?
2. Can the man be caught in, on, or between objects?
3. Can he slip or trip? Can he fall on the same level or to another?
4. Can he strain himself by pushing, pulling, or lifting?
5. Is the environment hazardous (toxic gas, vapor, mist, fume, or dust, heat or radiation)? (See discussion in Chapter 6.)

Close observation and job knowledge are required. The job observation can be repeated as often as necessary until all hazards and potential accidents have been identified.

Include hazards that might result. Record the type of accident and the agent involved. To note that a man might injure a foot by dropping a fire extinguisher, for example, write down "struck by extinguisher."

Again check with the observed employee after the hazards and potential accidents have been recorded. The experienced employee will probably suggest additional ideas. You should also check with others experienced with the job. Through observation and discussion, you will develop a reliable list of hazards and potential accidents.

Develop solutions

The final step in a JSA is to develop a recommended safe job procedure to prevent occurrence of potential accidents. The principal solutions are:

1. Find a new way to do the job.

JOB ANALYSIS WORK SHEET

JOB: Using a Pressurized Water Fire Extinguisher

WHAT TO DO (*Steps in sequence*)	HOW TO DO IT (*Instructions*)	KEY POINTS (*Items to be emphasized*) *Safety is always* *a key point*
1. Remove extinguisher from wall bracket.	1. Left hand on bottom lip, fingers curled around lip, palm up. Right hand on carrying handle palm down, fingers around carrying handle only.	1. Check air pressure to make certain extinguisher is charged. Stand close to extinguisher, pull straight out. *Have firm grip, to prevent dropping on feet.* Lower, and as you do remove left hand from lip.
2. Carry to fire.	2. Carry in right hand, upright position.	2. Extinguisher should hang down alongside leg. (This makes it easy to carry and reduces possibility of strain.)
3. Remove pin.	3. Set extinguisher down in upright position. Place left hand on top of extinguisher, pull out pin with right hand.	3. Hold extinguisher steady with left hand. Do not exert pressure on discharge lever as you remove pin.
4. Squeeze discharge lever.	4. Place right hand over carrying handle with fingers curled around operating lever handle while grasping discharge hose near nozzle with left hand.	4. Have firm grip on handle to steady extinguisher.
5. Apply water stream to fire.	5. Direct water stream at base of fire.	5. Work from side to side or around fire. After extinguishing flames, play water on smouldering or glowing surfaces.
6. Return Extinguisher. Report Use.		

JOB SAFETY ANALYSIS TRAINING GUIDE

DEPARTMENT: SV - SHOP

REQUIRED AND/OR RECOMMENDED PERSONAL PROTECTIVE EQUIPMENT: SAFETY GLASSES, HAND & FOOT PROTECTION

JOB: VERTICAL LATHE OPERATION

TITLE OF MAN WHO DOES JOB: LATHE MACHINIST   FOREMAN/SUPR:

SECTION: MACHINE SHOP

DATE: AUGUST 29, 19 --

ANALYSIS BY:

REVIEWED BY:

APPROVED BY:

SEQUENCE OF BASIC JOB STEPS	POTENTIAL ACCIDENTS OR HAZARDS	RECOMMENDED SAFE JOB PROCEDURE
1. PLACE RAW STOCK BLANK IN LATHE CHUCK.	1. (SA) Sharp edges of stock blank 2. (CB) Stock and lathe 3. (SO) Lifting raw stock blank 4. (CO) Moving parts while rotating blank 5. (SB) Falling Objects	1. Wear gloves or protect against sharp edges as required when handling raw stock. 2. Watch position of hands, arms and body to keep clear of pinch points. 3. Use proper lifting procedures. Get help if blank is too large for one man. 4. Keep hands clear when rotating chuck to position part. Use jogging control or slow speed. Wear short sleeves. Do not wear gloves or rings. 5. Wear foot protection
2. FABRICATE AND/OR INSTALL TEMPLATE ON STYLUS PLATEN.	1. (F) Slips on oily surface at same or different level. Climbing for access to elevated controls of equipment.	1. Keep oil spills wiped up. Apply non-skid material to elevated steps. Use work platform engineered for job, including hand rail protection.
3. SELECT AND INSTALL TOOL CUTTER.	1. (SB) Falling objects, oily 2. (SA) Sharp tools	1. Keep parts free of oil when handling, wear foot protection. 2. Be aware of handling sharp objects.
4. SELECT MACHINE MODE	1. (CW) Electrical controls	1. Be sure machine is properly grounded and all electrical controls are in good repair.
5. MACHINE PART	1. (SB) Metal particles	1. Wear eye protection.
6. REMOVE TURNINGS AND CHIPS AS MACHINING PROGRESSES	1. (SB) Turnings and chips 2. (CO) Turnings (More)	1. Wear eye protection. Use correct tools to clean turnings from table. 2. Break turnings before they become unwieldy. Do not wear long sleeve work clothes.

A completed JSA should become the basic guide for instructing employees. In this training guide, the abbreviations in the middle column are as follows: CB—caught between; CO—caught on; CW—contact with; F—falls; SA—struck against; SB—struck by; and SO—strained on.

2. Change the physical conditions that create the hazards.
3. To eliminate hazards still present, change the job procedure.
4. Try to reduce the necessity of doing a job, or at least the frequency that it must be performed. This is particularly helpful in maintenace.

To find an entirely new way to do a job, determine the work goal of the job, and then analyze the various ways of reaching this goal to see which way is safest. Consider work-saving tools and equipment.

If a new way cannot be found, then ask this question about each hazard and potential accident listed: "What change in physical condition (such as change in tools, materials, equipment, or location) will eliminate the hazard or prevent the accident?"

When a change is found, study it carefully to find what other benefits (such as greater production or time saving) will accrue. These benefits should be pointed out when proposing the change to higher management. They make good selling points.

The third step in solving the job-hazard problem is to investigate changes in the job procedure. Ask of each hazard and potential accident listed: "What should the man do—or not do—to eliminate this particular hazard or prevent this potential accident?" Where appropriate, ask an additional question, "How should he do it?" In most cases, the supervisor can answer these questions from his own experience.

Answers must be specific and concrete if new procedures are to be any good. General precautions—"be alert," "use caution," or "be careful"—are useless. Answers should precisely state what to do and how to do it. This recommendation—"Make certain the wrench does not slip or cause loss of balance"—is only "half good." It does not tell how to prevent the wrench from slipping.

Here, in contrast, is an example of a good recommended safe procedure that tells both "what" and "how": "Set wrench securely Test its grip by exerting a slight pressure on it. Brace yourself against something immovable, or take a solid stance with feet wide apart, before exerting full pressure. This prevents loss of balance if the wrench slips."

Often a repair or service job has to be repeated frequently because a condition needs correction again and again. To reduce the necessity of such a repetitive job, ask "What can be done to eliminate the cause of the condition that makes excessive repairs or service necessary?" If the cause cannot be eliminated, then ask "Can anything be done to minimize the effects of the condition?"

Machine parts, for example, may wear out quickly and require frequent replacement. Study of the problem may reveal excessive vibration is the culprit. After reducing or eliminating the vibration, the machine parts last longer and require less maintenance.

This fourth step—reducing frequency of a job—contributes to safety only in that it limits the exposure. Every effort still should be made to eliminate hazards and to prevent potential accidents through changing physical conditions or revising job procedures or both.

Finally, check or test the proposed changes by reobserving the job and discussing the changes with the men who do the job. Their ideas about the hazards and proposed solutions may be of considerable value. They can judge the practicality of proposed changes and perhaps suggest improvements. Actually these discussions are more than just a way to check a JSA. They are safety contacts that promote awareness of job hazards and safe procedures.

Benefits of JSA

The principal benefits that arise from job safety analysis are in these phases of a supervisor's work:

- Giving individual training in safe, efficient procedures.
- Making employee safety contacts.
- Instructing the new man on the job.
- Preparing for planned safety observations.
- Giving pre-job instructions on irregular jobs.
- Reviewing job procedures after accidents occur.
- Studying jobs for possible improvement in job methods.

Job Instruction Training

The job instruction training (JIT) method described here is often called the Four-Point Method, because the instructing job is broken into four parts, each of which will be discussed in detail:

1. Preparation,
2. Presentation,

3. Application, and

4. Testing.

The four-point method is intended to help an instructor teach a learner to do a specific job. It aims at faster learning and better learning. When combined with a JSA, it becomes an excellent method for teaching safety along with job skills. (The reader will be relieved to know that it takes longer to describe some of these steps than it does to do them.)

Step 1—Preparation

Decide what is going to be taught and how the major points will be presented.

Next, the instructor must get ready the necessary tools, equipment, and materials (including necessary safety equipment). Teaching can be ruined if it does not go off smoothly.

The learner, too, must be prepared by being "put at ease." If he is going to absorb instruction, if he is going to give full attention, he must not feel tense or self-conscious or worried.

Ways to put the learner at ease include: chatting with him, getting the learner to talk about himself and his experiences, having a cup of coffee together, telling of some amusing experience, or, in short, acting as though the instructor is glad to meet the fellow and anxious to get better acquainted. The instructor should be sincere about this, and if he cannot, chances are he will not like instructing and never be very good at it.

Create interest in the job. The instructor tells the learner what the purpose of the job is, how the job relates to other jobs in the plant or to the main business of his company, and other interesting facts about the job. He makes the person feel that learning the job will be a real accomplishment—something that he can be proud of and that will help him in the future. Nothing lets a person down as much as telling him, "There's really nothing to this job—anybody can learn it." Although this may be said in an attempt to give the learner confidence, it really makes him feel that he will hang on only as long as he has to while he looks around for a job that really amounts to something.

Instead of saying this, the instructor should tell the person that this job takes some learning but he is just the man to learn it, and that the instructor is going to stick with him to see that he does.

Relating the new job to something he has done before gives a man confidence and understanding. Careful observation and questioning will enable the instructor to know what experience the learner already has, so that the instructor can use actual examples. The instructor can also judge better when this first step should be brought to a close. Do not spend any more time on this step than necessary.

Step 2—Presentation

No more than five or six new points should be presented at one time. If a job consists of more than five or six parts, it can be organized into two or more units for instruction.

The instructor should both show and tell the learner how the job is done. He begins by doing the job himself—the "regular way" for two or three times, and then step by step, explaining what he is doing. He names parts, tools, and processes carefully and accurately. He uses key points as reminders: "until you can feel it strike against the gage," in a machine operation; "just tight enough to turn the ratchet," in using a micrometer; "until it reaches this mark," in reading a gage; "with your hand doubled into a fist," a safety key point in some operations; "walk your hands, don't slide them," a safety key point in handling lumber. The instructor may go through this phase of showing and telling several times slowly. He must use the same names and key points.

When the instructor shows and tells the learner how to do a job, the learner should observe from the approximate position of the operator. He can sit or stand by the instructor or he can look over his shoulder. If he faces the instructor, he will see the job backward and not adjust himself to the new situation as quickly.

Step 3—Application

Now the learner takes over and goes through the operation slowly, explaining what he is doing, using the correct names and key points taught in step 2. The instructor may prompt him when necessary, but does not touch the material or equipment. To overcome temptation to do this, the instructor might keep his hands in his pockets.

If, after prompting, the learner still is unable to go ahead on his own, the instructor goes back to step 2 and repeats the whole pro-

HOW TO GET READY TO INSTRUCT

Have a Timetable—
how much skill you expect him to have, by what date

Break Down the Job—*
list important steps. pick out the key points. (Safety is always a key point.)

Have Everything Ready—
the right equipment, materials and supplies.

Have the Workplace Properly Arranged—
just as the trainee will be expected to keep it.

*Use JSA, Job Safety Analysis breakdown to locate hazards.

JOB INSTRUCTION TRAINING (JIT)
HOW TO INSTRUCT

1. Prepare
Put trainee at ease.
Define the job and find out what he already knows about it.
Get him interested in learning job.
Place in correct position.

2. Present
Tell, show, and illustrate one IM-PORTANT STEP at a time.
Stress each KEY POINT.*

3. Try Out Performance
Have him do the job—coach him.
Have him explain each key point to you as he does the job again.
Make sure he understands.
Continue until YOU know HE knows.

4. Follow-Up
Put him on his own.
Designate to whom he goes for help.
Check frequently. Encourage questions.
Taper off extra coaching and close follow-up.

*Safety is always a key point.

SAFETY TRAINING INSTITUTE
NATIONAL SAFETY COUNCIL

Every supervisor should follow the JIT format when teaching job skills. Cards are available through NSC.

cedure while the learner again observes.

As soon as the learner can do the job without prompting, the instructor has him do it again to fix it in his mind. Then they proceed to step 4.

Step 4—Testing

The learner does the job, without explaining, while the instructor observes and corrects if necessary. Gradually the learner picks up speed, but not at the expense of perfect performance. If the learner does not seem sure or makes mistakes, the instructor asks him to do the job slowly and explains each step as he goes along (step 3). If he does a good job, the instructor compliments him.

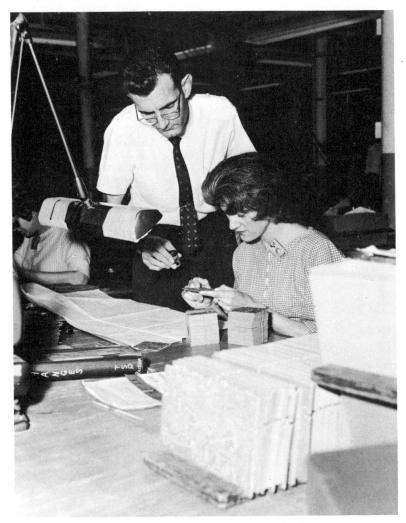

After an employee has been instructed in her job, the supervisor must check occasionally, not only to see how she is getting along, but also to check her performance.

Followup

After the steps of the four-point method of job instruction training (JIT) have been completed, the instructor tells the learner to keep on turning out good work, but not yet try for speed. The instructor asks

Showing actual items of equipment is a good training technique.

him to come back for help if he needs it, and from time to time, the instructor stops back to observe and comment on progress. He checks quality and quantity and watches closely to make sure that the learner is not "slipping up" on necessary steps in the operation, particularly safe practices.

Neglect of safe practices does not always show up in the finished product. It is discovered only through observation of the man at work. If the learner is taking the necessary precautions, the instructor can well compliment him in a friendly manner, "That's the right way. You don't want to get hurt just when you're getting good on this job."

Other Methods of Instruction

The methods discussed so far have primarily been methods that are commonly used to teach job skills in on-the-job situations. However, there may be times when the supervisor will need to know how to make a presentation or teach in a classroom situation. He may be

101

asked to teach a lesson that is part of an overall course, such as one designed to train employees to become welders or pipe fitters. There may be other occasions when the supervisor finds it advisable to call his group together to explain a new procedure or method. At such times, the lesson plan format will be useful.

The lesson plan

The supervisor should be familiar with lesson plans. They are the blueprint for presenting material contained in a course outline or for presenting a single unit of instruction. In addition to standardizing training, lesson plans help the instructor.

1. Present material in proper order.
2. Emphasize material in relation to its importance.
3. Avoid omission of essential material.
4. Run his classes on schedule.
5. Provide for trainee (student) participation.
6. Increase his confidence, especially if he is new.

Names for the parts of a lesson plan may vary; even the order may not always be the same. The following is a good example of arrangement for a lesson plan:

1. Title: Must indicate clearly and concisely the subject matter to be taught.

2. Objective:
   a) Should state what the trainee should know or be able to do at the end of the training period.
   b) Should limit the subject matter.
   c) Should be specific.
   d) May be divided into a major and several minor objectives for each session.

3. Training aids: Should include such items as actual equipment or tools to be used, and charts, slides, films, etc.

4. Introduction:
   a) Should give the scope of the subject.
   b) Should tell the value of the subject.
   c) Should stimulate thinking on the subject.

When CCTV is used for training in job skills, make sure the camera is positioned so the picture shows the operation from the same angle as viewed by a student standing at the side of the instructor.

5. Presentation:
   a) Should give the plan of action.
   b) Should indicate the method of teaching to be used (lecture, demonstration, class discussion, or a combination of these).
   c) Should contain suggested directions for instructor activity ("Show chart," "Write key words on chalkboard").

6. Application: Should indicate, by example, how trainees will apply this material immediately (problems may be worked; a job may be performed; trainees may be questioned on understanding and procedures).

7. Summary:
   a) Should restate main points.
   b) Shoud tie up loose ends.
   c) Should strengthen weak spots in instruction.

8. Test: Tests help determine if objectives have been reached. They should be announced to the class at the beginning of the session.

9. Assignment: Should give references to be checked or indicate materials to be prepared for future lessons.

Programmed instruction

Programmed instruction may be used as a supplement to classroom and on-job training. Using self-contained teaching materials (including so-called "teaching machines"), programmed instruction permits the trainee to set an individual pace and to absorb knowledge in easy-to-take bits. The learning process is reinforced by requiring the trainee to answer questions and correct his own errors before progressing with the course.

There are a number of programmed instruction courses available in such areas as safety training, vocational training, and communications; many of these courses use multi-media materials, such as tapes, slides, films, and teaching machines with devices.

A complete list of courses and devices is available from the National Society for Programmed Instruction, P.O. Box 137, Cardinal Station, Washington, D.C. 20017.

Independent study

Courses offered through correspondence schools are called home study courses, or independent study courses.

Independent study courses combine the fundamentals of good training, guidance and counseling of a qualified instructor, with the convenience of studying at home or in supervised study sessions arranged by the company (often on company time).

A list of subjects taught by accredited private home study schools may be obtained from the National Home Study Council, 1601 18th Street, NW., Washington, D.C. 20009.

Extension programs of most major universities offer programs of all types through independent study and will furnish complete information upon written request.

Closed circuit TV

Closed circuit TV (CCTV) training uses television's "instant replay" techniques. The basic technique is to record the visual procedure and directions on videotape and then play it back later on a monitor. Once the process or manual skills, along with the instructions are recorded, it can be replayed many times. It can be used to record the

104

steps in a job safety analysis, which can then be used to train employees in safe job procedures.

CCTV is extremely flexible; portions of the tape may be shown for review purposes where the job is not regularly performed. It also makes for uniform training since the job is presented in exactly the same manner to each trainee.

Summary

A number of training methods have been discussed in this chapter. Training may not solve all your problems, but good training will help solve many of them. The good supervisor attempts to fit the training method to the problem and not use the same technique to solve all problems. In many cases, the supervisor may find it advantageous to mix training methods, such as using programmed instruction along with on-the-job training to reach the highest level of learning in the shortest possible time.

Good training produces skilled and productive workers, as well as safe workers. Such workers make the supervisor's job easier.

CHAPTER 6

Industrial Hygiene

In addition to his safety responsibilities, each supervisor must make sure that there are no conditions in his work area that are detrimental to health. Consequently, the more he knows about industrial hygiene, the better supervisor he will be. This chapter should supply enough information to enable the first-line supervisor to recognize an environmental health hazard so that he can request assistance. Industrial hygienists, working with medical, safety, and engineering personnel can supply that help.

The industrial hygienist defines his work as "the recognition, evaluation, and control of environmental conditions that may have adverse effects on health, that may be very uncomfortable or irritating, or that may have some undesired effect upon the ability of an individual to perform his normal work."

It is possible to group these environmental conditions or stresses into four general categories—chemical, physical, ergonomic, and biological. So that the supervisor can recognize potential industrial hygiene problems, each will be discussed in some detail.

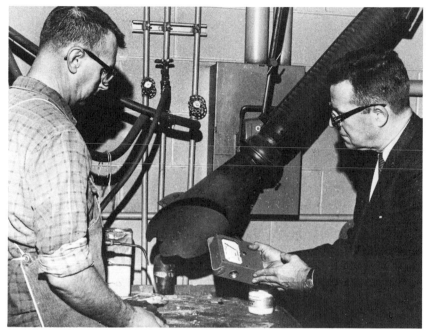

A velometer is used to measure air flow and obtain test data needed in connection with the proper function and design of industrial exhaust systems. In industry, harmful materials most commonly enter the body by inhalation of contaminated air.

Chemical

The first category is chemical. Chemical compounds in the form of liquids, gases, mists, dusts, fumes, and vapors may cause problems by inhalation (breathing), by absorbtion (through direct contact with the skin), or by ingestion (eating or drinking).

Inhalation

The major exposure to chemical compounds results from their being breathed after becoming airborne. Contaminants that can be inhaled into the lungs can be physically classified as gases, vapors, and particulate matter. Particulate matter can be further classified as a dust, fume, smoke, aerosol, or mist.

Absorbtion

Absorbtion through the skin can occur quite rapidly if the skin is cut or abraded; intact skin, however, offers a reasonably good barrier to chemicals. Unfortunately, there are several compounds that either in liquid or gaseous form, or both, may be absorbed through intact skin. Some are absorbed by way of the hair follicles and others dissolve in the fats and oils of the skin.

Examples of chemical compounds that can be hazardous by skin absorbtion are alkaloids; phenols; lead acetate; lead oleate; salts of lead, antimony, arsenic, bismuth, and mercury; nitrobenzene; nitrotoluene; aniline, and nitroglycerine. Other bad actors are triorthocresylphosphate, tetraethyl lead, and parathion and related organic phosphates. Compounds that are good solvents for fats may cause problems by being absorbed through the skin, although they are not as hazardous as those mentioned previously. Benzol, toluol, and xylol are examples of such solvents.

As a rule, most electrolytes and water are not a problem.

Ingestion

Ordinarily, people do not knowingly eat or drink harmful chemicals. It is true that toxic compounds capable of being absorbed from the gastrointestinal tract into the blood—an example is lead oxide—can cause serious problems if people working with them are allowed to eat or smoke in their work areas. Also, careful and thorough wash-ups are required before eating and at the end of every shift.

Physical Classification of Airborne Materials

Because inhalation of airborne compounds or materials is so easy, it is necessary that the supervisor have some idea of their physical classifications.

Dusts

These are solid particles generated by handling, crushing, grinding, rapid impact, detonation, and decrepitation (breaking apart by heating) of organic or inorganic materials, such as rock, ore, metal, coal, wood, and grain. Dusts do not tend to flocculate (aggregate) except under electrostatic forces; they do not diffuse in air, but settle under the influence of gravity.

Dust is a term used in industry to describe airborne solid particles that range in size from 0.1 to 25μ (1μ = 1/10,000 cm = 1/25,000 in.; μ is the abbreviation for micron). Dusts above 5μ in size usually will not remain airborne long enough to present an inhalation problem.

Dust may enter the air from various sources. It may be dispersed when a dusty material is handled, such as when lead oxide is dumped into a mixer or when talc is dusted on a product. When solid materials are reduced to small sizes in processes such as grinding, crushing, blasting, shaking, and drilling, the mechanical action of the grinding or shaking device supplies energy to disperse the dust formed.

Fumes

Solid particles generated by condensation from the gaseous state, generally after volatilization from molten metals, are called fumes. This physical change is often accompanied by a chemical reaction, such as oxidation. Fumes flocculate (aggregate loosely) and sometimes coalesce (fuse or grow together).

A fume is formed when a volatilized solid, such as a metal, condenses in cool air. The solid particles that make up a fume are extremely fine—usually less than 1.0μ. In most cases, the hot material reacts with the air to form an oxide. Examples are lead oxide fume from smelting, and iron oxide fume from arc welding. A fume also can be formed when a material such as magnesium metal is burned or when welding or gas cutting is done on galvanized metal.

Gases and vapors are *not* fumes, even if newspaper reporters often (and incorrectly) call them that.

Smoke

Carbon or soot particles less than 0.1μ in size, which result from the incomplete combustion of carbonaceous materials such as coal or oil, are called smoke. Smoke generally contains droplets as well as dry particles. Tobacco, for instance, produces a wet smoke composed of minute tarry droplets. The size of the particles contained in tobacco smoke is about 0.25μ.

Aerosols

Liquid droplets or solid particles dispersed in air, that are of fine enough particle size to remain so dispersed for a period of time, are called aerosols.

Mists

Mists are suspended liquid droplets generated by condensation from the gaseous to the liquid state or by breaking up a liquid into a dispersed state, such as by splashing, foaming, or atomizing. Mist is formed when a finely divided liquid is suspended in the atmosphere. Examples are the oil mist produced during cutting and grinding operations, acid mists from electroplating, acid or alkali mists from pickling operations, paint spray mist from spraying operations, and the condensation of water vapor to form a fog or rain.

Gases

Normally gases are formless fluids which occupy the space or enclosure in which they are confined and which can be changed to the liquid or solid state only by the combined effect of increased pressure and decreased temperature. Gases diffuse. Examples are welding gases, internal combustion engine exhaust gases, and air.

Vapors

The gaseous form of substances that are normally in the solid or liquid state (at room temperature and pressure) are called vapors. The vapor can be changed back to the solid or liquid state either by increasing the pressure or decreasing the temperature alone. Vapors also diffuse. Evaporation is the process by which a liquid is changed into the vapor state and mixed with the surrounding atmosphere. Solvents with low boiling points will volatilize readily.

Hazards involved

The hazard associated with breathing a gas, vapor, or mist usually depends upon the solubility of the substance that is airborne. For example, if the compound is very soluble—such as ammonia, formalde-

110

hyde, sulfuric acid, and hydrochloric acid—it is rapidly absorbed in the upper respiratory tract and does not penetrate deeply into the lungs. Consequently the nose and throat become so irritated that a person is driven out of the exposure area before he is in much danger from the toxicity of the gas. Nevertheless, exposures, even for brief periods, to high concentrations of these compounds can cause serious problems.

Compounds that are insoluble in body fluids cause considerably less pain than the soluble ones, and can penetrate deeply into the lungs. Thus, a very serious hazard can be present and not be immediately recognized. Examples of such gases are nitrogen dioxide and ozone. The immediate danger from these compounds in high concentrations is acute edema or, possibly later, pneumonia or circulatory impairment.

There are numerous chemical compounds that do not follow the general solubility rule. They are not very soluble in water and yet are very irritating to the eyes and respiratory tract. They also can cause lung damage and even death under the right conditions. An example is acrolein.

Solvents

The widespread industrial use of solvents can cause major problems. The supervisor can assist the hygienist and safety professional in helping to maintain control over exposure. Substantial exposures, fortunately, are usually controlled—spray-painting booths are ventilated and degreasing tanks are exhausted. These do not, ordinarily, present serious hazards as long as everything goes well. It is the job of the supervisor to make sure that controls and personal protective equipment are properly maintained and are used.

Small exposures—jobs that come up infrequently or which involve small amounts of solvents not covered by standard operating procedures—pose the most problems. The point to remember is not how much solvent is used at the jobsite, but how much solvent a person can inhale. A close check must be kept on all minor uses of solvents. A solvent should be issued only after it has been determined that it is to be used properly and safely. Industrial health authorities or your safety professional has information as to what solvents are to be used for what purpose, and how they are to be used. It is up to the supervisor to see that this information is used and that men abide by the recommendations.

Selection and handling

Getting the job done without hazard to employees or property is dependent upon the proper selection, application, handling and control of solvents and an understanding of their properties.

The term "solvent" refers to those organic liquids commonly used to dissolve other organic materials. It includes materials such as naphtha, mineral spirits, gasoline, turpentine, benzene, alcohol, and trichloroethylene.

A good working knowledge of the physical properties, nomenclature, and effects of exposure is very helpful in making a proper assessment of a solvent exposure.

Nomenclature can be misleading. For example, "benzine" is sometimes referred to by the worker as "benzene," a completely different solvent. Some commercial grades of benzine may contain benzene as a contaminant.

The supervisor should study the labels placed on containers by the manufacturer giving the specific name and composition of the solvents involved. Labels should tell what precautions are necessary when the solvents are used. Information indicating health and fire hazards is extremely important. In fact, if a solvent is not properly labeled, it should not be used. The purchasing department should notify suppliers that only properly labeled solvents will be accepted in the plant.

Degree of severity of solvent hazards

The severity of hazard in the use of organic solvents depends on the following factors:

1. How the solvent is used,
2. Type of job operation (determines how the workers are exposed),
3. Work pattern,
4. Duration of exposure,
5. Operating temperature,
6. Exposed liquid surface,

METHANOL

DANGER! FLAMMABLE
VAPOR HARMFUL
MAY BE FATAL OR CAUSE BLINDNESS
IF SWALLOWED
CANNOT BE MADE NONPOISONOUS

Keep away from heat, sparks, and open flame.
Keep container closed.
Avoid prolonged or repeated breathing of vapor.
Use only with adequate ventilation.

 **POISON**

Call A Physician
First Aid

If swallowed: Give a tablespoonful of salt in a glass of warm water
and repeat until vomit fluid is clear. Give two teaspoonfuls of
baking soda in a glass of water. Have patient lie down and keep
warm. Cover eyes to exclude light.

METHANOL MIXTURES

For products containing methanol in proportion sufficient to create hazard because
of methanol content, use applicable statements as above, with addition of:

CONTAINS OVER.................% OF METHANOL

The word "FLAMMABLE" may be omitted from the Statement of Hazards if the
product has a flash point above 80°F.

MCA Chemical Safety Data Sheet SD-22 available

Hazardous chemical label recommended by the Manufacturing Chemists'
Association.

7. Ventilation efficiency,
8. Evaporation rate of solvent,
9. Pattern of air flow,
10. Concentration of vapor in workroom air,
11. Housekeeping.

The solvent hazard therefore is determined not only by the toxicity
of the solvent itself but by the conditions of its use—who, what, how,
where, and how long.

For convenience, job operations employing solvents may be divided into three categories:

1. Direct contact is a consequence of hand operation. Emergency repair of equipment, spraying or packaging volatile materials without ventilation, cleanup of spills, and manual cleaning using cloths or brushes wetted with solvent are examples where employees may directly contact the solvent.

2. Intermittent or infrequent contact is encountered where the solvent is contained in a semi-closed system where exposure can be controlled. Paint spraying in an exhausted spray booth, vapor degreasing in a tank with local lateral slot exhaust ventilation, charging reactors or kettles in a batch type operation, and transferring liquids to secondary containers are examples of operations where the worker is exposed only at infrequent intervals.

3. Minimal contact is characterized by remote operation of equipment totally isolated from the work area. This type of operation includes directing chemical plant operations from a control room, mechanical handling of bulk packaged materials, and other operations where the solvent is contained in a closed system and is not discharged to the atmosphere in the work area.

"Safety" solvents. It is unfortunate that the term "safety" solvent has been applied to some proprietary cold cleaners, because the term is not precise and is subject to various interpretations. For example, a "safety" solvent may be considered by some users as nondamaging to the surfaces being cleaned. Other users may consider it to be free from fire or toxicity hazards. Depending on the conditions of use, neither of these criteria may be met by a so-called "safety" solvent.

These solvents are prepared as mixtures of halogenated hydrocarbons and petroleum hydrocarbons to be used for cold cleaning. Although the halogenated hydrocarbons are very effective grease and oil solvents and generally have no flash or fire point, they are relatively expensive and may be toxic under adverse conditions. The petroleum hydrocarbons are effective solvents, low in toxicity, and inexpensive, but they have flash and fire points.

To combine the best qualities of each solvent, manufacturers mix them in an attempt to produce a cold cleaner that has a flash point higher than that of the flammable petroleum hydrocarbon. Such a

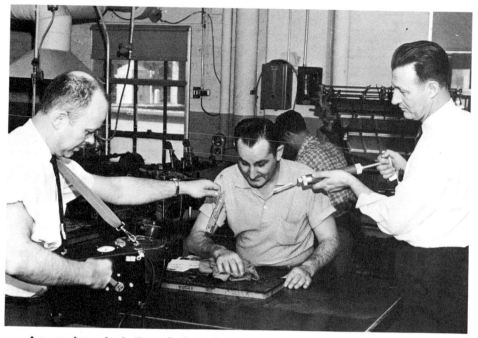

A supervisor who believes he has a dust, fume, vapor, or radiation hazard should consult an industrial hygienist. Here lab experts, using a madget impinger and an indicator tube sampler, check air for presence of solvent vapor in breathing zone.

mixture, however, can present both fire and toxicity hazards, depending on the evaporation rates of the solvents used; if the flammable liquid is more volatile than the nonflammable solvent, the vapors from the mixture can be highly flammable; conversely, if the nonflammable solvent is more volatile, it can evaporate to leave a flammable liquid.

These considerations should always be kept in mind by users of so-called "safety" solvents.

Particulates

Dusts and particulate matter such as fumes also fall in the chemical category.

To properly evaluate dust exposures requires knowledge of the chemical composition, particle size, dust concentration in air, method

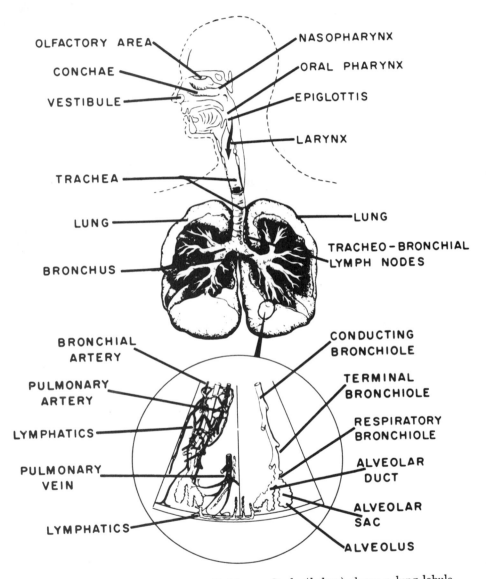

OLFACTORY AREA

CONCHAE

VESTIBULE

NASOPHARYNX

ORAL PHARYNX

EPIGLOTTIS

LARYNX

TRACHEA

LUNG

BRONCHUS

LUNG

TRACHEO-BRONCHIAL
LYMPH NODES

BRONCHIAL
ARTERY

PULMONARY
ARTERY

LYMPHATICS

PULMONARY
VEIN

LYMPHATICS

CONDUCTING
BRONCHIOLE

TERMINAL
BRONCHIOLE

RESPIRATORY
BRONCHIOLE

ALVEOLAR
DUCT

ALVEOLAR
SAC

ALVEOLUS

Cutaway illustration of the bronchial tree.   Circle (below) shows a lung lobule greatly enlarged.

116

of dispersion, and many other factors described in this section.

With the exception of some fibrous materials, dust particles must usually be smaller than 5μ in order to enter the alveoli or inner recess of the lungs. See the cutaway illustration of the bronchial tree. Although a few particles up to 10μ in size may enter the lungs, occasionally, nearly all the larger particles are trapped in the nasal passages, throat, larynx, trachea, and bronchi, from which they are expectorated or swallowed into the digestive tract.

A person with normal eyesight can detect dust particles as small as 50μ in diameter. Smaller airborne particles can be detected individually by the naked eye only when strong light is reflected from them. Dust of respirable size (below 10μ) can be seen only with the aid of a microscope. Most industrial dusts consist of particles that vary widely in size, with the small particles greatly outnumbering the large ones. Consequently, with few exceptions, when dust can be seen in the air around an operation, probably more invisible dust particles than visible ones are present.

Dispersion

In order for particulate matter to become airborne, some form of energy is required. A solid or liquid particle of sufficient mass will be thrown a considerable distance if ejected from its source with a high enough velocity. This type of dispersion is known as dynamic projection and is a result of the kinetic energy of the particle's motion. As the mass of the particle decreases, however, a point will be reached when its kinetic energy (which is one-half the mass times the square of the velocity) is too small to overcome air resistance. The particle's forward velocity is thus minimized and it remains suspended in the containing air mass.

As a rough approximation, macroscopic particles (those large enough to be visible to the naked eye) are considered to be dispersed by dynamic projection. Microscopic particles (those visible only through a microscope) are considered to have a mass so small that their movement is dependent on the containing air mass. Contaminants such as the larger dust particles, mists, and sprays, which are dispersed by dynamic projection, can cause external injury such as acid burns, eye damage, and dermatitis. The microscopic particles may be dangerous to health if inhaled.

117

Local exhaust pickup is placed so that particles from grinding operation are thrown directly into it.

Nonmetallic dusts

A number of occupational diseases result from exposures to non-metallic dusts, such as silica and asbestos dusts. Although specialized knowledge and instruments are needed to tell how severe a hazard may be, the supervisor can, however, recognize a danger spot and ask for expert help. If a supervisor is not alert, a lot of harm can be done.

Here are some of the things to watch for.

A process that produces dust fine enough to remain suspended in the air long enough to be breathed should be regarded as questionable until it can be proved safe. Silicosis, for example, is produced in hard rock mines and by quarrying and dressing granite, which contains free silica.

The danger of lung disease from the grinding wheel itself disappeared when the natural sandstone wheel was replaced with the Carborundum wheel and the Alundum wheel. Neither produces silicosis;

however, when they are used to grind castings that contain mold sand, the grinding operation can still present a silicosis hazard.

Methods of drilling rock with power machinery produce more dust than old-fashioned hand methods. This dust, however, is controlled by applying water to the drill steel so that the dust forms a slurry instead of forming an air suspension.

Processes in which inorganic materials are crushed, ground, or transported are potential sources of dust. They should either be controlled by use of wet methods or should be enclosed and ventilated by local exhaust. Points where conveyors are loaded or discharged, transfer points along the conveying system, and heads or boots of elevators should be enclosed and, usually, exhaust ventilated. Powdered material that is simply riding along on a belt or other conveyor should cause no dust problem.

The supervisor must check that someone does not cancel the effectiveness of built-in dust controls by tampering with them or by using them improperly. He must also check closely that required respiratory equipment is worn by those workers who need supplementary protection.

Fumes

Welding, metalizing, and other hot operations produce fumes, which may be harmful under certain conditions.

Arc welding volatilizes metal which condenses—as the metal or its oxide—in the air around the arc. In addition, the rod coating is in part volatilized. These fumes, because they are extremely fine, are readily inhaled.

Iron pigment in the lungs appears to have no effect in producing illness or disability. Its presence shows up in X rays of the lungs and may lead to a mistaken diagnosis of silicosis.

More-toxic materials—such as those formed when welding structures that have been painted with red lead, or when welding galvanized metal—may produce severe symptoms of toxicity ·rather rapidly, unless fumes are controlled with good local exhaust ventilation, or the welder is protected by respiratory protective equipment.

When pouring brass, zinc volatilizes from the molten mass and oxidizes in the surrounding air to produce a zinc fume, which (in high concentrations) may produce the rather nonspecific disease known as

Soldering fumes are pulled away from operator by duct (upper right).

"metal fume fever" or "brass ague." If lead is present it also will become airborne. As a result, brass foundry workers may incur a lead intoxication—chronic if the lead is present only in very small amounts and serious if it is present in substantial amounts.

Most soldering operations, fortunately, do not require temperatures high enough to volatilize an appreciable amount of lead. However, some of the lead in the molten solder is oxidized by contact with air at the surface. If this oxide, often called dross, is mechanically dispersed into the air, it may produce a severe lead poisoning hazard.

In operations where this might happen, such as soldering or lead battery making, prevention of occupational poisoning is largely a matter of scrupulously clean housekeeping to prevent the lead oxide from even becoming dispersed into the air. It is customary to enclose melting pots, dross boxes, and similar operations, and to ventilate them strongly in order to control the hazard.

Physical Agents

The second category of environmental factors or stresses mentioned in the definition of industrial hygiene involves physical agents. Problems relating to such things as noise, ionizing radiation, nonionizing radiation, temperature extremes, and pressure extremes fall in this category. It is important that the supervisor be alert to the hazards because they may have immediate or cumulative effects on the health of the employees he supervises.

Noise

Noise—defined as "unwanted sound"—is a form of vibration which may be conducted through solids, liquids, or gases. The effects of noise on man include the following:

- Psychological effects, for example: noise can startle, annoy, and disrupt concentration, sleep, or relaxation.
- Interference with communication by speech, and as a consequence, interference with job performance and safety.
- Physiological effects, for example: noise-induced loss of hearing, or aural pain, even nausea (when the exposure is severe).

Damage risk criteria. If the ear is subjected to high levels of noise for a sufficient period of time, some loss of hearing may occur. A number of factors can influence the effect of the noise exposure. Among these are:

Variation in individual susceptibility;
The total energy of the sound;
The frequency distribution of the sound;
Other characteristics of the noise exposure, such as whether it is continuous, intermittent, or made up of a series of impacts;
The total daily time of exposure;
The length of employment in the noisy environment.

Because of the complex relationships of noise and exposure time to partial loss of hearing and the many other possible contributory causes for hearing loss, it is difficult to give exact rules for protecting workers against hearing loss. However, criteria have been developed to pro-

121

tect against hearing loss in the speech frequency range. In fact, these criteria, known as the Threshold Limit Values for noise, are observed by the U.S. Department of Labor under the Occupational Safety and Health Act; they have the force of law. See Table III.

TABLE III
PERMISSIBLE EXPOSURES

Duration per Day Hours	Sound Level dB(A)*
8	90
6	92
4	95
3	97
2	100
1½	102
1	105
¾	107
½	110
¼	115-C**

* Sound level in decibels as measured on a standard level meter operating on the A-weighting network with slow meter response.
** Ceiling Value

Limits have been set for exposure to noise, but, normally, it will not be possible for the supervisor to measure noise levels. This should be done by an industrial hygienist or a safety professional who then gives the supervisor the necessary data for establishing administrative controls or advises as to the necessity for hearing protection. Details and definitions of noise and its measurement, evaluation, and control are given in NSC's *Accident Prevention Manual for Industrial Operations, Fundamentals of Industrial Hygiene,* and *Basics of Industrial Noise and Audiometry.*

Rules of thumb. There are three non-technical "rules of thumb" that the supervisor can use to determine if his work area has excessive noise.

1. If it is necessary to speak in a very loud voice or shout directly into the ear of a person in order to be understood, it is likely that the

exposure limit for noise is being exceeded.

2. If employees tell you that they have head noises and ringing noises in their ears at the end of the work day, they are being exposed to too much noise.

3. If employees complain that speech or music sounds muffled to them after leaving work but sounds fairly clear in the morning when they return to work, there is no doubt about their being exposed to noise levels that can eventually cause a partial loss of hearing that can be permanent.

The supervisor can prevent excessive exposure to noise by working closely with engineering personnel in attempting to reduce noise levels. If the noise cannot be reduced by engineering means, the supervisor can resort to administrative controls, whereby he rotates employees so that the time they spend in a noisy area does not exceed a specified limit. If this is not feasible or if engineering controls can not be installed for a long period of time, the supervisor must enforce the wearing of ear protection devices. In this respect, as with all aspects of safety, he must set a good example.

Ionizing radiation

Although ionizing radiation is a complex subject, a brief discussion is included here because it is one of the more important physical agents.

Gamma radiation from radioactive materials and X radiation are highly penetrating and may produce damage in any tissue of the body. What tissue will be damaged depends in part on the energy of the radiation and in part on the relative sensitivity of the tissue.

Alpha and beta radiation are generally also considered with gamma and X radiation, although they are not as penetrating and are usually warded off by the skin. Their danger arises when the materials which produce them have been ingested and fixed in some tissue where the radiations can carry on their destruction at close range.

In the handling of some types of radioactive materials and in the operation of the high-powered particle accelerators now being used, bombardment by high-speed protons and by both high-speed and thermal neutrons must also be considered as a possible hazard to health.

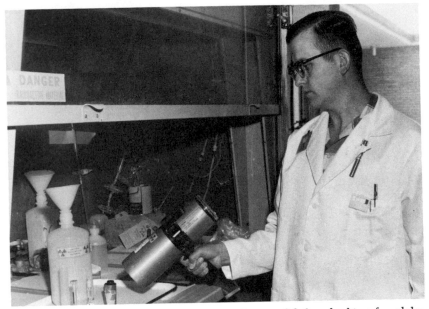

Ionization chamber rate meter ("cutie-pie") is useful for checking for alpha, beta, and gamma radiation. Note dosimeter on man's coat lapel.

These types of radiation all cause injury by ionizing the tissue in which they are absorbed. Such injuries differ widely in locations and extent but the determining factors are the ability of the particular radiation to penetrate through the tissues and the amount of ionization produced by a given physical amount of the radiation. Alpha radiation, for instance, is classed as 20 times as biologically effective as X ray because the alpha particles are heavy and possess a great deal of ability to disrupt and ionize the tissue molecules with which they come in contact.

X radiation is produced by high-potential electrical discharge in a vacuum and should be anticipated in evacuated electrical apparatus operating at a potential of 10,000 volts or more. Although X radiation is certain to be produced in such apparatus, it is dangerous only if it penetrates the jacket of the apparatus and enters the inhabited part of the workroom.

If one is being overexposed to the radiation of a comparatively low-voltage X-ray tube, an X-ray dermatitis of the hand is generally the first result to be seen. It is characterized by a rough, dry skin, a wartlike growth, and dry, brittle nails. With continuing exposure and somewhat more penetrating X ray, bone destruction will develop.

Although this discussion is concerned primarily with X rays and gamma radiation because they are the best known, the same statements apply qualitatively to other types of penetrating radiation. The general method for preventing injuries from any penetrating radiations is separation from them by distance or by shielding with heavy substances in sufficient quantity to reduce the radiation received to below the maximum permissible dose and by limiting the time of exposure.

Nonionizing radiation

Electromagnetic radiation has varying effects on the body, depending largely on the particular wavelength of the radiation involved. Following, in approximate order of decreasing wavelength and increasing frequency, are some hazards associated with different regions of the nonionizing electromagnetic radiation spectrum.

Low frequency. The longer wavelengths—including power frequencies, broadcast radio, short wave radio—can produce general heating of the body. The health hazard from these radiations is very small, however, since it is unlikely that they would be found in intensities great enough to cause significant effect.

Microwaves have wavelengths of 3 m to 3 mm (100 to 100,000 mega Hertz, MHz). They are found in radar, communications, and diathermy applications. Microwave intensities may be sufficient to cause significant heating of tissues.

Effect is related to wavelength, power intensity, and time of exposure. Generally, the longer wavelengths will produce a greater temperature rise in deeper tissues than the shorter wavelengths. However for a given power intensity there is less subjective awareness to the heat from longer wavelengths than there is to the heat from shorter wavelengths because of its absorption beneath the body's surface.

An intolerable rise in body temperature, as well as localized damage, can result from an exposure of sufficient intensity and time. In addition, flammable gases and vapors may ignite when they are inside metallic objects located in a microwave beam.

Power intensities for microwaves are given in units of watts per sq cm. Areas having a power intensity of over 0.01 watt per sq cm

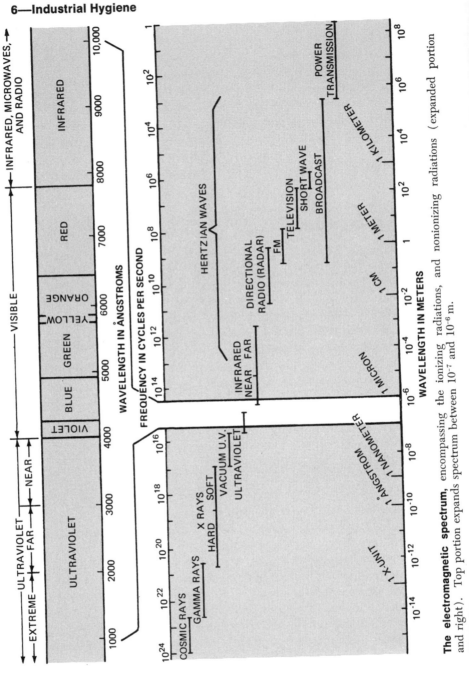

The electromagnetic spectrum, encompassing the ionizing radiations, and nonionizing radiations (expanded portion and right). Top portion expands spectrum between 10^{-7} and 10^{-6} m.

should be avoided. In such areas, dummy loads should be used to absorb the energy output while equipment is being operated or tested. If a dummy load cannot be used, adjacent populated areas should be protected by adequate shielding.

Infrared radiation does not penetrate below the superficial layer of the skin so that its only effect is to heat the skin and the tissues immediately below it. Except for thermal burns, the health hazard upon exposure to conventional infrared radiation sources is negligible.

Visible radiation and ultraviolet radiation also do not penetrate appreciably below the skin; their effects are essentially heating on the

Good lighting and good housekeeping are engineered into this machine shop.

surface. The effects of the ultraviolet waves are much more violent than are those of the visible or infrared—a severe burn can be produced with no warning whatever and significant damage to the lens of the eye can occur from excessive exposure.

Since ultraviolet radiation in industry may be found around electrical arcs, they should be shielded by materials opaque to the ultraviolet. Opacity in the ultraviolet has no relation to opacity in the visible part of the spectrum. Ordinary window glass, for instance, is almost completely opaque to the ultraviolet although transparent to the visible. A piece of plastic dyed a deep red-violet may be almost entirely opaque in the visible part of the spectrum and transparent in the near ultraviolet.

Electric welding arcs and germicidal lamps are the most common strong producers of ultraviolet in industry. The ordinary fluorescent lamp generates a good deal of ultraviolet inside the bulb, but it is essentially absorbed in the glass bulb and its fluorescent coating.

The most common exposure to ultraviolet radiation is from direct sunshine. Men who continually work outdoors in the full light of the sun may develop tumors on exposed areas of the skin. These tumors occasionally turn malignant.

Ultraviolet radiation from the sun also increases the skin effects of some industrial materials. After exposure to compounds such as cresols, the skin is exceptionally sensitive to the sun. Even a short exposure in the late afternoon when the sun is low is likely to produce a severe sunburn. There are other compounds which minimize the effect of UV-rays. Some of these are used in certain protective creams.

Visible Radiation

Visible radiation, which falls about halfway up the spectrum of frequency and wavelength, concerns the supervisor because it can effect both the quality and accuracy of work. Good lighting invariably results in increased product quality with less spoilage and increased production. Good lighting contributes to sanitary, clean, and neat operations.

What is good lighting? Many factors help produce it: color of the light, direction and diffusion, and amount. The nature of illuminated surfaces is important. A dark gray, dirty surface may reflect only 10 or 12 percent of the incident light, while a light-colored, clean surface may reflect more than 90 percent.

Lighting should be bright enough to produce easy seeing and directed so that it does not create glare. The level should be high enough to permit efficient seeing. The *Accident Prevention Manual for Industrial Operations* lists the illumination levels and brightness ratios recommended for manufacturing and service industries by the

Illuminating Engineering Society.*

One of the most objectionable features of lighting is glare—brightness within the field of vision that causes discomfort or interferes with seeing. The brightness can be caused by either direct or reflected light. To prevent glare, keep the source of light well above the line of vision, or shield it with opaque or translucent material.

Objectionable glare can originate from a surface that reflects the image of the light source directly into the line of vision, such as can polished control handles or machinery dials. You can dull reflections by having such surfaces matte (dull), rather than polished or smooth.

Almost as bad as this is an area of excessively high brightness in the visual field. A highly reflecting white paper in the center of a dark, nonreflecting surface, or a brightly illuminated control handle, or dial, on a dark or dirty machine are two examples.

To prevent this condition, keep surfaces light or dark with little difference in surface reflectivity. Color contrasts are all right, however.

Although it is generally best to provide even, shadow-free light, some jobs require contrast lighting. In these cases, keep the general, or background, light well diffused and glareless, and add a supplementary source of light that is directed to cast shadows as wanted. Be careful not to have too much contrast in brightness between the work and its surroundings, and not to have direct or reflected glare.

Lasers

Lasers emit beams of coherent light of a single color (or wavelength). Contrast this with conventional light sources that produce random, disordered light wave mixtures of various wavelengths. The word laser is an acronym for Light Amplification by Stimulated Emission of Radiation. A laser is made up of light waves that are nearly parallel to each other (collimated) and are all traveling in the same direction. Light-emitting atoms are "pumped" full of energy and stimulated to fall to a lower energy level, giving off light waves that are directed to produce the coherent laser beam.

The maser, the laser's predecessor, emits microwaves instead of

* Illuminating Engineering Society, *Practice for Industrial Lighting*, is designated as American National Standard A11.1-1965 (R1970).

light. Some companies call their lasers "optical masers." Proposed uses for the laser include machining and cutting of metals, welding of microscopic parts, and systems for high-capacity communications.

Since the laser is highly collimated (has a small divergence angle), it can have a large energy density in a narrow beam. Direct viewing of the laser should therefore be avoided. The work area should contain no reflective surfaces (such as mirrors or highly polished furniture) for even a reflected laser beam can be hazardous. Suitable shielding to contain the laser beam should be provided.

Biological effects. The eye is the organ most vulnerable to injury induced by laser energy. The reason for this is the ability of the cornea and lens to focus the parallel laser beam on a small spot on the retina.

The fact that infrared radiation of certain lasers may not be visible to the naked eye contributes to the potential hazard.

Lasers generating in the ultraviolet range of the electromagnetic spectrum produce corneal burns rather than retinal damage, because of the way the eye handles ultraviolet light.

Other factors which have a bearing on the degree of eye injury induced by laser light are: (a) the pupil size—the smaller the pupil diameter, the less the amount of laser energy permitted to the retina, (b) the power of the cornea and lens to focus the incident light on the retina, (c) the distance from the source of energy to the retina, (d) the energy and wavelength of the laser, (e) the pigmentation of the subject, (f) the place on the retina where the light is focused, (g) the divergence of the laser light, and (h) the presence of scattering media in the light path.

Temperature Extremes

General experience shows that extremes of temperature affect the amount of work that a man can do and the manner in which he does it. The industrial problem is more often that of exposure to high temperatures rather than low temperatures.

The body is continuously producing heat through its metabolic processes. Since the body processes are so designed that they can operate only within a very narrow limit of temperature, they must dissipate this heat as rapidly as it is produced if the body is to function efficiently and well. A sensitive and rapidly acting set of thermostatic devices in the body must also control the rates of its temperature-regulating processes.

Sweating

Sweating is the most important of the temperature-regulating and heat-dissipating processes. Almost all parts of the skin are provided with sweat glands which excrete a liquid (mostly water and a little salt) to the surface of the body. This goes on continuously, even under conditions of rest.

In an individual who is resting and not under stress, the sweating rate is approximately 1 L (liter) per day, which is evaporated as rapidly as it is excreted. Under the stress of heavy work or high temperature, this sweating rate may increase to as much as 4 L (approximately one gallon) per 4 hr. As much as 10 to 12 gm (150 to 190 grains) of salt per day will be lost with the water. Both the water and salt loss must be made up rather promptly if good health is to continue. In terms of stress, an individual's ability to acclimate is variable and such stress is dangerous for those with heart problems.

If the sweat can be evaporated as rapidly as it can be formed and if the heat stress is not such as to exceed the maximum sweating rate of which the body is capable, the body can maintain the necessary constant temperature.

The rate of evaporation depends upon the moisture content, the rate of air movement, and the temperature of the surrounding air. The industrial hygienist must simultaneously deal with all of these factors in any attempt to determine the effect of the thermal environment on man.

Radiation

Radiation accounts for some of the body's thermal equilibrium with its surroundings. If an object in the surroundings is far below body temperature, such as a large glass window on a subzero day, a large amount of heat can be radiated and the person may feel chilled even if the air in his immediate environment is fairly warm. Conversely, if an object in the surroundings, such as a furnace wall, is very much above body temperature, a person may receive a large amount of heat by radiation, and it may be extremely difficult to keep him cool enough by the other available means to maintain a constant body temperature.

Radiant heat is electromagnetic energy that does no heating whatsoever until it strikes some object, like a person, where it is absorbed. No relief from it whatsoever results from blowing air around. The only protection is to set up an invisible infrared "shadow." This can be done by placing any kind of opaque shield or screen between the man and the radiating surface, such as discussed later under "Heat controls."

Conduction

Conduction through the clothing and dissipation into the air by conduction are the source of some cooling to the body. This usually is not an important means of cooling because the conductivity of clothing and the heat capacity of air are usually low.

Conduction and convection become an important means of heat loss when the body is in contact with a good cooling agent, such as water. For this reason, when people are exposed to cold water, they become chilled much more rapidly and effectively than when exposed to air at the same temperature.

Air movement cools the body by conduction and convection, but, more important, moving air removes from about the body the film of saturated air (which is formed very rapidly by evaporation of sweat) and replaces it with a fresh layer, capable of accepting more moisture.

Effects of high temperature

Effects of high temperature are counteracted by the body's attempt to keep the internal temperature down by increasing the rate of heart beat. The capillaries in the skin then dilate to bring more blood to the surface so that both the rate of cooling and, gradually, the body temperature are increased.

If the thermal environment is tolerable, these measures will rather soon lead to an equilibrium where the heart rate and the body temperature remain constant. If this equilibrium is not reached until the body temperature is about 102 F, corresponding to a sweating rate of about 2 L per hr, there is imminent danger of heatstroke.

Intermittent rest periods for persons necessarily exposed to ex-

Open hearth worker opens a tap hole. He is protected by a long coat and hood made of asbestos-polyester cloth; gloves are sewn to cuffs of the coat. Worker must be fully protected not only from the heat, but also from premature breakout or splash of molten steel.

treme heat reduces this danger.

Heatstroke (also known as sunstroke) is not necessarily the result of exposure to the sun. It is caused by exposure to an environment in which the body is unable to cool itself sufficiently. As a result, the body temperature rises, and reaches a point where the heat-regulating mechanism breaks down completely. The body temperature then rises rapidly.

The symptoms are hot dry skin, severe headache, visual disturbances, rapid temperature rise, and loss of consciousness. The condition is recognizable by the flushed face and high temperature. The victim should be removed from the heat immediately and his body cooled as rapidly as possible, most readily by being wrapped in cool wet sheets. Since the condition may be quickly fatal, medical help should be obtained as soon as possible.

Heatstroke is a much more serious condition than heat cramps or heat exhaustion. An important predisposing factor is excessive physical exertion. The only method of control is to reduce the temperature of the surroundings or to increase the ability of the body to cool itself, so that body temperature does not rise. Heat shields, discussed under "Heat Controls," next, are of value here.

Heat cramps may result from exposure to high temperature for a relatively long time, particularly if accompanied by heavy exertion, with excessive loss of salt and moisture from the body. Even if the moisture is replaced by drinking plenty of water, an excessive loss of salt may provoke heat cramps or heat exhaustion.

Heat cramps are characterized by the cramping of the muscles of either the skeletal system or the intestines. In either case the condition may be relieved in a few hours under proper treatment, although soreness may persist for several days.

Heat exhaustion may result from physical exertion in a hot environment when vasomotor control and cardiac output are inadequate to meet the increased demand placed upon them by peripheral vasodilation. Its signs are a relatively low temperature, pallor, weak pulse, dizziness, profuse sweating, and a cool moist skin.

Prevention of heat exhaustion and heat cramps has become almost routine since the introduction of the use of salt tablets in the hot occupations. There is some resistance to the use of salt tablets on the part of a few individuals who say that they are nauseated by them. Usually, nausea results when the tablets are taken with an inadequate amount of water to dilute the salt. Generally, an adequate salt supply is provided by a well balanced diet.

To prevent nausea, a full 8-oz glass of water should be taken with each tablet. It can also be prevented by the use of enteric-coated tablets (which do not dissolve until they have passed the stomach) or by the use of tablets impregnated with cellulose which dissolve very slowly in the stomach contents. Industrial experience with a salted drinking water supply has been good. A salt concentration in the drinking water of ½ gm/L (0.05 percent) is tasteless and supplies all the salt necessary to replace the loss caused by exposure to heat.

Experience has shown that men do not stand a hot job very well at first, but develop tolerance rapidly and acquire full acclimatization within a week to a month.

Heat controls

A great deal can be done in "hot" industries to improve the health, efficiency, and comfort of workers exposed to high temperature and humidity. Both equipment and good planning are required. The supervisor must intelligently use those facilities which may be subject, in part at least, to his control.

Heat shields, as well as ventilation, may also be used. A large radiant area may heat a person excessively, even if the air surrounding the person is not very hot. In these cases, it is more effective to provide a reflector to return the radiant energy toward its source than it would be to try to cool the person by air blast or to try to insulate the radiant source.

Sheet aluminum reflectors completely surrounding, but not touching, a furnace will reflect 95 percent (or more) of the radiant energy, and will themselves absorb so little heat that they will be cooled by natural circulation of the air around them. Not only will these reflectors completely protect workmen from excessive radiation, they will also reduce the atmospheric temperature because they decrease the amount of heat received and re-radiated by other objects.

The supervisor should not attempt to use makeshift remedies like asbestos or darkened metal shields. A rusty iron panel, for example, will absorb radiation from the source, become heated, and then act as a radiator itself. Fortunately, aluminum foil, sheets, and corrugated siding are all highly reflective in the infrared—and they are inexpensive.

Atmospheric Pressures

It has been recognized from the beginning of caisson work, dating back to about 1850, that men working under greater-than-normal atmospheric pressures are subject to various ills connected with the job. The main effect (decompression sickness, commonly known as the "bends") results from the release of nitrogen bubbles into the circulation and tissues during decompression. The bubbles lodge at the joints and under muscles, causing severe cramps. To prevent this trouble, decompression is carried out slowly and by stages so that the nitrogen can be eliminated slowly, without the formation of bubbles.

Watch where you locate a portable air compressor. Recirculation of contaminated air into the intake of an air supply that is intended to provide quality air for breathing is a common problem.

Deep-sea divers are supplied with a mixture of helium and oxygen for breathing. Since helium is an inert diluent and is less soluble in blood and tissue than is nitrogen, it presents a less formidable decompression problem.

Under some conditions of work at high pressure, the concentration of carbon dioxide in the atmosphere may be considerably increased so that the carbon dioxide will act as a narcotic. Keeping the oxygen concentration high will minimize the condition although not prevent it. This procedure is useful where the carbon dioxide concentration cannot be kept at a proper level.

One of the most common troubles encountered by workers under compressed air is pain and congestion in the ears from inability to ventilate the middle ear properly during compression and decompression. As a result, many workers under compressed air suffer from temporary and some from permanent loss of hearing. The cause of this damage is considered to be obstruction of the eustachian tubes which prevents proper equalization of pressure on the middle ear.

The effects of reduced pressure on the worker are much the same as the effects of decompression from a high pressure. If pressure is reduced too rapidly, decompression sickness and ear disturbances similar to, if not identical with, the diver's conditions may result.

Men working at reduced pressure are also subject to oxygen starvation, which can have very serious and insidious effects upon the senses and judgment. There is a considerable amount of evidence that exposure for 3 to 4 hr to an altitude of 9000 ft above sea level without breathing an atmosphere enriched in oxygen can result in severely impaired judgment. Even if pure oxygen is provided, the altitude should be limited to that giving the same partial pressure of oxygen as air at 8000 ft.

Reduced pressure is not the only condition under which oxygen starvation may occur. Deficiency of oxygen in the atmosphere of confined spaces is commonly experienced in industry. For this reason, the oxygen content of any tank or other confined space should be checked before entry is made. Instruments such as the oxygen analyzer are commercially available for this purpose.

Normal air contains approximately 21 percent oxygen by volume. The first physiologic signs of a deficiency of oxygen (anoxia) are increased rate and depth of breathing. Oxygen concentrations of less than 16 percent by volume cause dizziness, rapid heart beat, and headache. One should never enter or remain in areas where tests have indicated such low concentrations unless he is wearing supplied-air or self-contained respiratory equipment.

Oxygen-deficient atmospheres may cause inability to move and a semiconscious lack of concern about the imminence of death. In cases of sudden entry into areas containing little or no oxygen, the individual usually has no warning symptoms, but immediately loses consciousness and has no recollection of the incident if he is rescued and revived.

Atmospheric contaminants are discussed in a following section, "Threshold Limit Values" and "Standard operating procedures."

Ergonomics

The third category of environmental factors can be called ergonomic. It involves human reaction to monotony, fatigue, repeated motion, and repeated shock.

The term "ergonomics" literally means the customs, habits, and laws of work. According to the International Labor Office it is:

". . . The application of human biological science in conjunction with the engineering sciences to achieve the optimum mutual adjustment of man and his work, the benefits being measured in terms of human efficiency and well-being."

The ergonomics approach goes beyond productivity, health, and safety. It includes consideration of the total physiological and psychological demands of the job upon the workers.

In the broad sense, the benefits which can be expected from designing work systems to minimize physical stress on workers are:

> More efficient operation,
> Fewer accidents,
> Lower cost of operation,
> Reduced training time, and
> More effective use of personnel.

The human body can endure considerable discomfort and stress and can perform many awkward and unnatural movements—for a limited period of time. When, however, unnatural conditions or motions are continued for prolonged periods, the physiological limitations of the worker may be exceeded. To ensure a continuingly high level of performance, work systems must be tailored to human capacities and limitations.

Biomechanics is that phase of engineering devoted to the improvement of the man-machine-task relationship in an effort to reduce operator discomfort and fatigue.

Biotechnology is a broader term. It encompasses biomechanics, human factors engineering, and engineering psychology. To arrive at biotechnological solutions to work-stress problems, the sciences of anatomy, physiology, psychology, anthropometry, and kinesiology need to be brought into play.

Program areas of concern are:

- Strictly biomechanical aspects—the consideration of stress on muscles, bones, nerves, and joints.
- Sensory aspects—the consideration of eye fatigue, color, audio signals, and the like.
- External environment—such as lighting, glare, temperature, humidity, noise, atmospheric contaminants, and vibration.
- The psychological and social aspects of the working environment.

Information obtained from studying these factors can be translated into tangible changes in work environments. Among improvements that are reducing fatigue and stress are redesigned hand tools, adjustable chairs and workbenches, improvements in lighting, control of heat and humidity, and noise reduction.

Mechanical vibration

A condition known to stonecutters as "dead fingers" or "white fingers," occurs mainly in the fingers of the hand used to guide the cutting tool. The circulation in this hand becomes impaired, and when exposed to cold the fingers become white and lose sensation, as though mildly frostbitten. The condition usually disappears when the fingers are warmed for some time, but a few cases are sufficiently disabling that the men are forced to seek other types of work. In some instances, both hands are affected.

The condition has been observed in a number of other occupations involving the use of fairly light vibrating tools, such as the air hammers used for scarfing and chipping in the metal trades. The condition is produced by vibration while the fingers are held in a strained position and is at least aggravated by chilling of the fingers at the same time. Prevention of this condition is much more satisfactory than treatment.

Preventive measures include directing the exhaust air from air-driven tools away from the hands so that they will not become unduly chilled, use of handles of a comfortable size for the fingers and, in some instances, substitution of mechanical cleaning methods for some of the hand methods which have produced most of the cases of "white fingers." In many instances, simply preventing the fingers from becoming chilled while at work will eliminate the condition.

Repeated motion

Repetitive motions or repeated shocks like those in sorting and assembling jobs often cause irritation and inflammation of the tendon sheath of the hands and arms. The condition is generally known as tenosynovitis, and when once established, it is painful and disabling. The usual treatment is rest and diathermy.

The condition results from excessive strain many times repeated. It may occur to an employee who has been working at the same job for years if he is suddenly asked to put in considerable overtime. It is most likely to occur to a new employee or to an employee transferred to a new job. Prevention of the condition is, of course, much more satisfactory than treatment.

For example, in the printing industry cases of sore hands and wrists were prevented by lowering the surface of the jogging tables 2 in. so that the hands and wrists could fall into a relaxed position at the end of each motion.

The employees should be closely watched so that if they show signs of soreness in the backs of the hands, wrists, forearms, or shoulders they can be transferred temporarily to other work or the job can be changed slightly to reduce the strain.

Biological Stresses

The fourth and final category of environmental factors is of a biological nature and is related to problems with animal handling, insects, molds, fungi, bacteria, and grain dusts.

Tuberculosis

Tuberculosis and other infections may be an occupational disease when they are contracted by nurses, doctors, or attendants caring for tuberculosis patients, or during autopsy or laboratory work where the bacilli may be present.

Fungus infections

A number of occupational infections are common to workers in agriculture and in closely related industrial jobs. Grain handlers who inhale grain dust are likely to come in contact with some of the fungi which contaminate grain from time to time. Some of these fungi can and do flourish in the human lung, causing a condition called farmer's lung.

Fine, easily spread fungi, such as rust and smut, can be contacted once the covering of infected grain is broken and may produce sensitization. Infections from some of these fungi are endemic (of local nature) in the Southwest and are not necessarily industrial.

Byssinosis

Byssinosis occurs in individuals who have experienced prolonged exposure to heavy air concentrations of cotton dust. Flax dust has also been incriminated.

The exact mode of action of the cotton dust is unknown, but one or more of these factors may be important: (a) toxic action of microorganisms adherent to the inhaled fibers, (b) mechanical irritation from the fibers, and (c) allergic stimulation by the inhaled cotton fibers or adherent materials.

It takes several years of exposure before manifestations are noticed.

Anthrax

Anthrax is a highly virulent bacterial infection. In spite of considerable effort in quarantining infected animals and in sterilizing of imported animal products, this disease remains a problem. With prompt detection and modern methods of treatment, it is much less likely to be fatal, as it was a few years ago.

"Q fever"

There have been reports of infection—"Q fever"—with a rickettsial organism among meat and livestock handlers. It is similar to but apparently not identical with "tick fever," which has been known for many years. The exact mode of transmission is not known, but there is evidence that "Q fever" may come from contacting freshly killed carcasses or droppings of infected cattle. Probably in the latter case transmission is by inhalation of the infectious dust. Prevention undoubtedly depends primarily upon recognition and elimination of the disease in the animal host.

Brucellosis

Brucellosis (undulant fever) has long been known as an infection produced by drinking unpasteurized milk from cows suffering from Bang's disease (infectious abortion). Since it can also be contracted by handling the animals or their flesh and is also transmitted by swine and goats, it is an occupational ailment of slaughterhouse workers as well as of farmers, although more common among the latter. Here also, preventive measures are primarily proper testing and control of the animals to eradicate the disease.

Erysipelas

The other major bacterial infection to which slaughterhouse workers and fish handlers are subject is erysipelas, which seems to be especially virulent when contracted from the slime of fish.

Upper respiratory tract infections

Workers exposed to dusts from vegetable fibers such as cotton, bagasse (sugar cane residues), hemp, flax, or grain may develop upper respiratory tract infections ranging from chronic irritation of the nose and throat to bronchitis, complicated by asthma, emphysema, or pneumonia, or a combination.

Dusts from many of these fibers may also be a source of allergens, histamine, and toxic metabolic products of microorganisms. However, most of the cases of illness arising from handling these fibers have been reported in the foreign literature.

Threshold Limit Values

Chemical compounds cause problems only because they end up inside the body. An open container of carbon tetrachloride sitting on a work bench is not a problem until someone stands over it and breathes the vapor. Also, the carbon tetrachloride vapor shouldn't be a problem if diluted with enough air to keep it consistently below a concentration of 10 parts of carbon tetrachloride per million parts of air (ppm) in the breathing zone. Thus limits, called Threshold Limit Values (abbreviated as TLV), have been established for airborne concentrations of many chemical compounds. Since the supervisor could be involved in discussing problems with his employees and industrial hygienists who are conducting surveys, he should understand something about TLVs and the terminology in which their concentrations are expressed.

The basic idea of TLVs is fairly simple. They refer to airborne concentrations of substances and represent conditions under which it is believed that nearly all workers may be repeatedly exposed, day after day, without adverse effect. Because individual susceptibility varies widely, exposure of an occasional individual at (or even below) the threshold limit may not prevent discomfort, aggravation of a pre-existing condition, or occupational illness. In addition to the TLVs set for chemical compounds, there are limits for physical agents, such as noise, microwaves, and heat stress.

The TLV may be a time-weighted average figure which would be acceptable for an 8-hour exposure. For some substances, such as an extremely irritating one, a time-weighted average concentration would not be acceptable, so a "Ceiling" value is established. In other words, a Ceiling limit means that at no time during the 8-hour work period should the airborne concentration exceed the limit.

Establishment of values

Threshold Limit Values have been established for about four hundred substances. A group of well qualified toxicologists, hygienists, and doctors reviews the list annually and values are revised as necessary.

The data for establishing a TLV comes from animal studies, human studies, and industrial experience. The limit may be selected for one of several reasons. It may be based on the fact that a substance is very irritating to the majority of people exposed to concentrations above a given level. The substances may be an asphyxiant. Other reasons for establishing a limit might be that the chemical compound is anesthetic, or fibrogenic, or can cause allergic reactions, or malignancy. Some TLVs are established because above a certain airborne concentration, a nuisance exists.

The concentrations of airborne materials capable of causing problems are quite small. Consequently the hygienist uses special terminology to define these concentrations. He often talks in terms of parts per million (ppm) when describing the airborne concentration of a gas or vapor. If he is measuring airborne particulate matter, such as a dust or fume, he uses the term milligrams per cubic meter (mgs/M³) to define concentrations. He also may use the actual number of particles present in one cubic foot of air being tested, in which case the descriptive term becomes millions of particles per cubic foot (mppcf). As an example of the very small concentrations involved the hygienist commonly samples and measures substances in the air of the working environment in concentrations ranging from 1 to 100 ppm. Some idea of the magnitude of these concentrations can be appreciated when one realizes that 1 in. in 16 miles is a part per million, one cent in $10,000, one ounce of salt in 62,500 lb of sugar, one ounce of oil in 7812.5 gal of water—all represent one part per million.

Dermatitis

The four general categories into which industrial hygiene problems fall have been discussed in some detail, yet dermatitis has not been mentioned. Because skin problems in industry account for the single largest cost of any industrial hygiene problems, they are important enough to be placed in a category by themselves. The individual problems usually can be traced back to exposure to some kind of a chemical compound or to some form of physical abrasion or irritation of the skin. Occupational skin diseases account for about 60 percent of all compensation claims for occupational diseases. Although rarely a direct cause of death, skin disorders cause much discomfort and are often hard to cure.

Causes of occupational skin diseases are classified in these ways:

1. Mechanical agents—friction, pressure, trauma.
2. Physical agents—heat, cold, radiation.
3. Chemical agents—organic and inorganic. These are subdivided according to their action on the skin as primary irritants or sensitizers.
4. Plant poisons—several hundred plants and woods can cause dermatitis. The best known example is poison ivy.
5. Biological agents—bacteria, fungi, parasites.

144

Chloracne is caused by clothing that has been soaked in cutting oils. It could have been prevented by after-work showers and daily change of work clothes.

Even substances that are normally harmless will cause irritations of varying severity in some skins.

There are two general types of skin reaction: primary irritation dermatitis and sensitization dermatitis. Practically all persons suffer primary irritation dermatitis from mechanical, physical, or chemical agents. Brief contact with a high concentration of a primary irritant or prolonged exposure to a low concentration will result in inflammation. Allergy is not a factor in these conditions. Sensitization dermatitis, the second type reaction, is the result of an allergic reaction to a given substance. Once sensitization develops, even small amounts of the material may cause symptoms.

Some substances produce both types of dermatitis. Examples are organic solvents, formaldehyde, and chromic acid.

Occupational acne

Cutting fluids are frequently responsible for occupational acne. It all starts when the skin becomes irritated from contact with the oil.

Next a blackhead forms. If it becomes infected, it forms oil pimples or boils that resemble adolescent acne. Dirty oil, worker carelessness, and use of germicides in the fluid increase the possibility of occupational acne.

To reduce the possibility, keep machines and areas around them clean. Change cutting fluid at regular intervals. Have employees use protective creams or gloves, apron, or face shield. Encourage employees who work in these areas to shower at the end of each shift. They should use warm water, mild soap, and a soft brush.

Some types of skin are more susceptible than others, but most people will develop oil acne with sufficient exposure. Certain factors help account for some skin disorders. Any investigation of occupational skin disease should consider:

1. Degree of perspiration,
2. Personal habits of cleanliness,
3. Pre-existence of skin disorders and allergic states, and
4. Diet.

Some sensitive skins may require special cleansers.

Standard Operating Procedures

One of the primary functions of the supervisor is to assist in developing standard operating procedures for jobs that require industrial hygiene controls and to enforce these procedures once they have been developed. Common jobs involving maintenance and repair of systems for storing and transporting fluids or entering tanks or tunnels for cleaning and repairs are controlled almost entirely by the immediate supervisor.

Entering tanks

As an example, there should be a standard operating procedure for entering tanks. Even if a tank is empty, it may have been closed for some time and developed an oxygen deficiency. It is unsafe to enter without supplied air respiratory protection. If it has contained a solvent and has not been thoroughly cleaned, its atmosphere is saturated with the solvent vapor. Even if the volatile solvent has been

All confined areas, such as tanks and subsurface structures, must be tested for oxygen deficiency and toxic and flammable gases and vapors, and proper precautions taken before anyone is allowed to enter. Be sure to probe all areas.

flushed with water and steam, the tank may still be hazardous to enter. There are records of severe poisoning of workmen who have entered "cleaned" tanks that were found to be relatively free of solvent vapor. In these instances, the disturbed scale released trapped solvents.

For these reasons, it is universal practice to provide hose masks and blowers for men entering tanks. Because the inside of a washed tank looks relatively harmless, workmen often neglect to use the equipment supplied. It is the supervisor's job to make certain that no workman enters a tank without a hose mask and attached lifeline, nor without a man outside to assist him if he encounters serious difficulties.

147

Maintaining pumps and valves

The other job that can often lead to serious exposures to solvent vapors is maintenance of pumps, valves, and packed joints in solvent lines. Vapor from dripping packing glands may collect to form a fire and explosion hazard, as well as a health hazard.

The lower explosive limit for most flammable solvents is about $1\frac{1}{2}$ percent. The threshold limit value for less-toxic solvents is about $\frac{1}{10}$ of one percent. Control measures for solvent exposures try to keep the concentration below the threshold limit value for health reasons. Although the fire hazard will be taken care of "automatically" in general areas, a dripping gland could still produce a flammable concentration in its immediate area.

Industrial hygiene controls

Finally, the supervisor should be aware of the various kinds of controls that are used to maintain a good healthy working environment.

The type and extent of these controls depends upon the physical, chemical, and toxic properties of the air contaminant, the evaluation made of the exposure, and the operation that disperses the contaminant. The extensive controls needed for lead oxide dust, for example, would not be needed for limestone dust, since much greater quantities of limestone dust can be tolerated.

General methods of controlling harmful environmental factors or stresses include the following:

1. Substitution of a less harmful material for one which is dangerous to health.
2. Change or alteration of a process to minimize worker contact.
3. Isolation or enclosure of a process or work operation to reduce the number of persons exposed.
4. Wet methods to reduce generation of dust in operations such as mining and quarrying.
5. Local exhaust at the point of generation and dispersion of contaminants.
6. General or dilution ventilation with clean air to provide a safe atmosphere.
7. Personal protective devices, such as special clothing, and eye and respiratory protection. (These are discussed in the next chapter.)

This exhaust hood keeps solvent vapor out of the breathing zone of the employee.

8. Good housekeeping, including cleanliness of the workplace, waste disposal, adequate washing and eating facilities, healthful drinking water, and control of insects and rodents.
9. Special control methods for specific hazards, such as reduction of exposure time, film badges and similar monitoring devices, continuous sampling with preset alarms, and medical programs to detect intake of toxic materials.
10. Training and education to supplement engineering controls.

As a supervisor, you play a very important part in making sure that controls are effective. You must observe and study process changes, process enclosures, use of wet methods, local exhaust ventilation, and general ventilation very carefully and systematically to make sure that they are functioning properly or to determine if maintenance or repair work is needed. Even more important is the part you must play in controls relating to personal protective devices. The Occupational Safety and Health Act places the responsibility for enforcement of the wearing of such devices squarely upon supervision. Consequently if you do not enforce the use of ear protectors, for example, where there is no engineering way to reduce noise, you can be cited and your company penalized by the U.S. Department of Labor.

Good housekeeping, special monitoring devices, and training and education also call for the full cooperation of the supervisor.

In fact, you, the supervisor, are the key man as far as safety and industrial hygiene controls are concerned. Without your help and guidance, they will be of very little value.

CHAPTER7

Personal Protective Equipment

When a hazard is found to exist, every effort should be made to eliminate it or control it by engineering design. Where this is not possible, isolation of the process or the guarding of the hazard should be attempted. Personal protective equipment should be considered only as a last resort. Remember, the purpose of the Occupational Safety and Health Act of 1970 is "to assure as far as possible every working man and woman in the Nation safe and healthful working conditions. . . ."

Judicious use of personal protective equipment sometimes, however, can greatly simplify operations in case of breakdowns or other emergencies.

Controlling Hazards

Persons who must work where hazards cannot be eliminated or controlled at the source, and where ordinary work clothes do not afford sufficient protection, should use personal protective equipment, which, if necessary, can protect the person from head to toe.

You should do the following, if you are to have an effective pro-

151

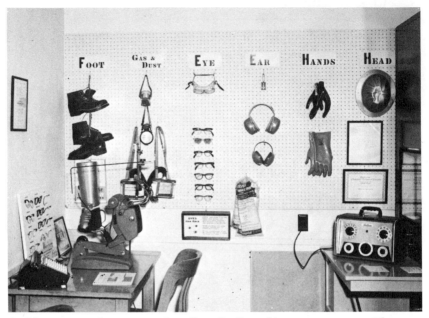

Accident prevention is more than "a policy" at this paper company. The most modern protective devices are made available at each mill (and other facility) for easy access by employees, because management feels that men are more likely to keep up their own personal safety gear supply.

gram dealing with safe dress and personal protective equipment.

- Be familiar with required standards and the Occupational Safety and Health Act requirements.
- Be able to recognize hazards.
- Be familiar with the best safety equipment available to protect against these hazards.
- Know the procedures for supplying the equipment.
- Know how to maintain and clean the equipment.
- Develop an effective method for persuading all employees to dress safely and to wear the proper protective equipment when they should.

In addition to knowing what personal protective equipment is needed and when it is needed, the supervisor is responsible for following up on the use of such equipment. When persons do not see the purpose of using protective equipment, the supervisor must get them to recognize the need for it. Getting some workers to use protective equipment may be one of the toughest jobs a supervisor must face.

To fulfill this responsibility, supervisors should learn about safety equipment. Local safety meetings are a good source of information because here equipment is usually displayed and manufacturers' representatives are handy. Hundreds of firms exhibit at the annual National Safety Congress and Exposition, held in Chicago each fall. Company safety and personnel departments can be helpful. There is a wealth of literature on safety equipment—manufacturers' catalogs, trade literature, and the "equipment issue" of *National Safety News*, published each March.

The safe rule to follow when specifying or buying all types of safety equipment is to insist on the best equipment and to deal with reputable firms. Do not take a chance on inferior items just because they may be less expensive. They should conform to standards where they apply. Respiratory protective equipment should be listed by NIOSH (National Institute of Occupational Safety and Health) or Bureau of Mines.

This chapter covers protection of the eyes and face, feet and legs, head, ears, fingers and hands, and the torso. Respiratory protective equipment, safety belts and harnesses, protection against ionizing radiation, and safe work clothing are discussed, as are methods of paying for certain equipment and getting employees to use it properly.

Protecting the Eyes

Industrial operations expose the eyes to a variety of hazards—flying objects, splashes of corrosive liquids or molten metals, dusts, and harmful radiation are common examples.

Eye injuries not only disable a person, but they often disfigure him. Per-injury cost is high to both employee and employer.

Flying objects cause most eye injuries—metal or stone chips, nails, or abrasive grits. The National Society for the Prevention of Blindness lists the chief causes of eye injury as:

Flying objects (especially those set in motion by hand tools)
Abrasive wheels (small flying particles)
Corrosive substances   Splashing metal
Injurious light or heat rays   Poisonous gas or fumes

153

Usually, in all operations where hardened metal tools are struck together, where equipment or material is struck by a metal hand tool, or where the cutting action of a tool causes particles to fly, eye protection is needed by the user of the tool and by other workmen who

A trained safety serviceman fits and adjusts safety glasses at employee work stations. In this job, worker needs front eye protection only.

may be exposed to flying particles. The hazard can be minimized through the use of nonferrous, "soft" striking tools and through shielding the job by metal, wood, or canvas.

Safety goggles or face shields should be worn when woodworking or cutting tools are used head-high or overhead with the chance of particles falling or flying into the eyes.

Occasionally, the need for eye protection is overlooked on such jobs as cutting wire and cable, striking wrenches, using hand drills, chipping concrete, removing nails from scrap lumber, shoveling material head-high or working on the leeward side of a job, and using wrenches and hammers overhead, and on other jobs where particles of materials or debris may fall.

154

Equipment types

Goggles and other eye protection equipment are available in many types and styles, and with the protective medium of heat- or chemically treated glass, plastic, wire screen, or light-filtering glass. Supervisors should be familiar with these various forms of eye protection and should know the best for the job. Protective equipment for the eyes includes:

Cover goggles
Protective spectacles
Spectacles with side shields
Cup goggles

Chemical goggles
Dust goggles
Miners' goggles
Melters' goggles
Welders' goggles

General information follows.

Cover goggles are frequently worn over ordinary spectacles. This protects both the wearer's eyes and his corrective lenses. Lenses that have not been heat or chemically treated are easily broken. A cover goggle protects against pitting as well as breaking.

Cover goggles include the cup type with heat-treated lenses and the wide-vision type with plastic lenses. Both types are used for heavy grinding, machining, chipping, riveting, work with molten metals, and similar heavy operations. They offer the advantage of being wide enough to protect the eye socket and distribute a blow over a wide area.

Tempered lenses are being more widely used in ordinary eye-corrective spectacles used for street wear, but many of them do not meet the requirements of American National Standard Z87.1–1968. The requirements of the Food and Drug Administration Standard,* effective December 31, 1971, are not as rigid as the ANSI Standard. Although they do provide additional protection against ordinary exposure, they are inadequate for industrial exposure; therefore, only lenses meeting the American National Standard should be worn for such exposures.

The wearer may have his correction ground in heat-treated lenses. A refractionist (ophthalmologist or optometrist) should do the fitting and should know the job on which eye protection is to be worn. He

* Food and Drug Administration, Dept. of Health, Education and Welfare, Arlington, Va. 22202.

Representative types of eye protectors. Upper left: metal frame safety spectacles. Upper right: plastic frame safety spectacles with side shields. Center: chipper's goggles for use over prescription spectacles. Lower left: ventilated rubber-frame monogoggle. Lower right: chemical splash protection goggle.

should also know the working distance, especially for those whose accommodation is not good (such as often happens in middle age) or who work at very close distances. Some companies require that only bifocal lenses be used in safety glasses that require a correction. This eliminates the risk of having a wearer lift his glasses up whenever he must see close up or far away, depending on his correction.

Protective spectacles without side shields may be worn where it is unlikely that particles will fly toward the side of the face, although spectacles with side shields are recommended for all industrial use. Side shield spectacles should always be worn where there is additional side exposure.

Frames must be rigid enough to hold lenses directly in front of the eyes. The nose bridge should be adjustable or universal. Frames should be fitted by a person trained for the job.

156

TABLE IV

FILTER LENS SHADE NUMBERS FOR VARIOUS
WELDING AND CUTTING OPERATIONS

Type of Operation	Shade Number
Shielded metal-arc welding—$\frac{1}{16}$-, $\frac{3}{32}$-, $\frac{1}{8}$-, $\frac{5}{32}$-inch electrodes	10
Gas-shielded arc welding (nonferrous)—$\frac{1}{16}$-, $\frac{3}{32}$-, $\frac{1}{8}$-, $\frac{5}{32}$-inch electrodes	11
Gas-shielded arc welding (ferrous)—$\frac{1}{16}$-, $\frac{3}{32}$-, $\frac{1}{8}$-, $\frac{5}{32}$-inch electrodes	12
Shielded metal-arc welding:	
$\frac{3}{16}$-, $\frac{7}{32}$-, $\frac{1}{4}$-inch electrodes	12
$\frac{5}{16}$-, $\frac{3}{8}$-inch electrodes	14
Atomic hydrogen welding	10–14
Carbon arc welding	14
Soldering	2
Torch brazing	3 or 4
Light cutting, up to 1 inch	3 or 4
Medium cutting, 1 inch to 6 inches	4 or 5
Heavy cutting, 6 inches and over	5 or 6
Gas welding (light) up to $\frac{1}{8}$ inch	4 or 5
Gas welding (medium) $\frac{1}{8}$ inch to $\frac{1}{2}$ inch	5 or 6
Gas welding (heavy) $\frac{1}{2}$ inch and over	6 or 8

Note: In gas welding or oxygen cutting where the torch produces a high yellow light, it is desirable to use a filter or lens that absorbs the yellow or sodium line in the visible light of the operation.

Notes:

1. All filter lenses and plates shall meet the test for transmission of radiant energy prescribed in ANSI Z87.1–1968, *Practice for Occupational and Educational Eye and Face Protection.*

2. All glass for lenses shall be tempered, substantially free from striae, air bubbles, waves, and other flaws. Except when a lens is ground to provide proper optical correction for defective vision, the front and rear surfaces of lenses and windows shall be smooth and parallel.

3. Lenses shall bear some permanent distinctive marking by which the source and shade may be readily identified.

Source: *U.S. Code of Federal Regulations,* Title 29, Chapter XVII, Part 1910, Occupational Safety and Health Standards, § 1910.252(e)(2)(i); and Part 1926, Safety and Health Regulations for Construction, § 1926.102(a)(5).

Chemical goggles, with soft vinyl or rubber frames, protect eyes against splashes of corrosive chemicals and exposure to fine dust or mist. Lenses can be of heat-treated glass or acid-resistant plastic. For exposures involving chemical splashes, they are equipped with baffled ventilators on the sides. For vapor or gas exposures, they must

157

TABLE V. EYE AND FACE PROTECTIVE EQUIPMENT TO BE WORN FOR VARIOUS TYPES OF EXPOSURES

	Heavy impact (chipping, grinding, etc.)	Light to moderate impact (belt sanding, most machining operations, etc.)	Frontal impact only, light to moderate	Dusts	Liquid splash, irritating only	Liquid splash, eye hazardous	Molten metal splash (melting point below 2,100 F)	Molten metal splash (melting point above 2,100 F)	Gases, fumes, mists
Spectacles (3mm glass or plastic lenses)									
" with sideshields									
Spectacles (0.050-inch plastic lenses)									
" with sideshields									
Eyecup goggles, basic (3mm lenses)									
" with fine-mesh screens									
" with shielded/baffled ports							High viscosity only	High viscosity only	
Rigid or semirigid mask goggles (3mm lens)									
" with fine mesh screens or baffled ports									
Rigid or semirigid mask goggles (0.050-inch lens)									
" with fine mesh screens or baffled ports									

Flexible-fitting mask goggles (3mm or 0.050-inch lens)

" with fine-mesh screens

" with shielded/baffled ports

" unventilated

Foundrymen's goggles (3mm lenses)

" with fine-mesh screens

Gas-tight goggles

Special fabric-cup dust goggles

Rubber-frame chemical goggles, baffled vents

" unventilated

Plastic-visor face shield (0.040-inch)

" with crown and chin protectors

Wire-mesh visor face shield (0.0295-inch mesh)

High viscosity only

High viscosity only

Recommended, Optimum Protection

Permissible

Permissible but has features not needed for hazard cited

159

be unventilated. Some types are made to fit over spectacles.

Leather mask dust goggles should be worn by employees who work around noncorrosive dusts, as in cement and flour mills. The goggles have heat-treated or filter lenses. Wire-screen ventilators around the eye cup provide air circulation.

Miners' goggles for underground work and other locations where fogging is a serious problem are made of a corrosion-resistant wire screen, coated a dull black to reduce reflection.

Melters' goggles come in spectacle and cup types with cobalt blue glass in graded shades. Lenses with color in the upper half and clear glass in the lower half are also available. Frames are of leather or plastic to protect the face from radiant heat.

Welders' goggles with filter lenses are available for such operations as oxyacetylene welding, cutting, lead burning, and brazing. On page 157 is a guide for the selection of the proper shade numbers. These recommendations may be varied to suit the individual's needs.

Goggle frames. All goggle frames should be of corrosion-resistant material that will neither irritate nor discolor the skin, that can withstand sterilization, and that are flame resistant or nonflammable. Metal frames should not be worn around electrical equipment or near intense heat.

To permit the widest range of vision, goggles should be fitted as close to the eyes as possible without the eyelashes touching the lens. The lenses should have no appreciable distortion or prism effect. Some specifications are covered by the American National Standard, Z87.1–1968, *Practice for Occupational and Educational Eye and Face Protection.*

Cleaning goggles. Goggles should be sterilized before they are reissued to other employees. The proper procedure is to disassemble and wash them with soap or detergent in warm water. Rinse thoroughly. Replace defective parts.* Immerse all parts in a solution containing germicide, deodorant, and a fungicide. Do not rinse, but hang to dry in air. Then place the parts in a clean, dustproof container.

Some companies keep extra goggles in the main supply department. Others keep them in each department, along with a stock of

---

* If a lens has more than just the most superficial scratch, nick, or pit, it should be replaced. Such damage can materially reduce protection afforded the wearer.

parts. A third practice is to have a goggle cart with a trained attendant who makes the rounds of the company to clean, adjust, repair, or replace goggles on the job. Many companies have stations which dispense cleaning liquid and tissues to encourage frequent cleaning as soon as possible after they get smudged or dirty.

Carrying cases or storage cabinets help promote proper care of goggles.

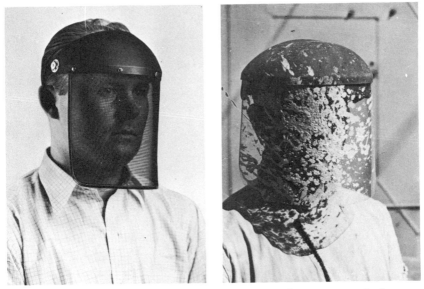

Left: metal-screen face shield that affords protection to furnacemen and others exposed to radiant heat. **Right:** when the entire face must be protected against light impact and chemical splash, a face shield should be provided. This shield stopped a splash of boiling caustic.

Face protection

Many types of personal protective equipment shield the face (and sometimes the head and neck also) against light impact, chemical or hot metal splashes, heat radiation, or other hazards.

Face shields of clear plastic protect eyes and face of a person who is sawing or buffing metal, sanding or light grinding, or handling chemicals. The shield should be slow burning and must be replaced when warped or scratched. A regular replacement schedule must be set up because plastic tends to become brittle with age.

Left: the lift front on this welding helmet permits the welder to inspect the work without raising the helmet. Stationary plate of clear impact-resistant glass affords protection for eyes if light chipping is required. **Right:** sturdy nonflammable helmet with specified cover and filter plates protects this welder against sparks, flying particles, and injurious rays.

The headgear and shield should be adjustable to the size and contour of the head, and be easy to clean. Many types permit the wearer to raise the shield without removing the headgear.

Metal screen face shields deflect heat away from a person and still permit good visibility. They are used around blast furnaces, soaking pits, heating furnaces, and other sources of radiant heat.

Babbitting helmets protect the head and face against splashes of hot metal, rather than against heat radiation. The helmet consists of a window made of extremely fine wire screen, a tilting support, an adjustable headgear, and a crown protector and chin protector (if required).

Welding helmets, shields, and goggles protect the eyes and face against the splashes of molten metal and the radiation produced by arc welding. Helmets should have the proper filter glass to keep ultraviolet and visible rays from harming the eyes. A list of shade numbers recommended for various operations is given in Table IV. Worker's eyes vary due to age, general health, and the care they give them. Two persons may require different shaded lenses when doing the same job.

Cracked or chipped filter lenses must be replaced, otherwise they will permit harmful rays to reach the welder's eyes.

The shell of the helmet must resist sparks, molten metal, and flying particles. It should be a poor heat conductor, and a nonconductor of electricity. Helmets that develop pinholes or cracks must be discarded. Helmets should have a headgear that permits the worker to use both hands and also to raise the helmet to position work.

Many types of helmets have a replaceable, heat-treated glass or plastic covering to protect the filter lens against pitting and scratching. Some helmets have a lift-front glass holder which permits the welder to give his work a rapid inspection without lifting or removing the helmet.

Impact goggles worn under the helmet protect a welder from flying particles when the helmet is raised. The spectacle type with side shields is recommended as minimum protection from flash from adjacent work or from popping scale of a fresh weld.

A welder's helper should have proper welding goggles or helmet to wear while assisting in a welding operation or while chipping flux away after a bead has been run over a joint. The danger of foreign bodies becoming lodged in the eye is great for workmen performing this latter operation without protection.

A hand-held shield can be used where the convenience of a helmet is not needed, such as for inspection work, tack welding, and other operations requiring little or no welding by the user. Frame and lens construction is similar to that of the helmet.

Welding goggles are available with filter glass shades up to No. 8. If darker shades are needed, then complete face protection is needed because of the danger of skin burns. When shades darker than No. 3 must be used, a side-shield or cup goggles is recommended.

Acid-proof hoods that cover head, face, and neck are used by persons exposed to severe splashes from corrosive chemicals. This type hood has a window of glass or plastic that is securely joined to the hood to prevent acid from seeping through. Hoods made of rubber, neoprene, plastic film, or impregnated fabrics are available for resistance to different chemicals. Consult the manufacturer to find the protective properties of each material.

Hoods with air supply should be worn for work around toxic fumes, dusts, gases, or mists. Because an unventilated hood quickly becomes warm and humid, an air line is added mainly for comfort. The worker should wear a harness or belt to support the hose.

163

Transparent face shields, supported from a head harness or headband, are used where there is limited exposure to direct splashes of corrosive chemicals. Splashproof cup goggles should be worn under the shield for added protection.

Overcoming employee complaints

Perhaps the most common complaint that people give as to why they do not want to wear eye protection is that it is uncomfortable. People cannot be expected to wear protection if it is uncomfortable. That is why equipment must be carefully fitted. Frames must be tight and straight and properly adjusted to the wearer. The eye size should be correct for the wearer, the bridge size and temple must likewise be right. Each pair of glasses should be adjusted to the wearer by a qualified person. Most goggles are light weight, barely heavier than eyeglasses, so their weight should present no problem.

Another complaint is that people say they cannot see with goggles. Dirty, dusty, grimy goggles do interfere with vision. The supervisor must instruct people not to touch lenses with dirty hands. Glasses should never be set down with the lenses down. They can be scratched or pitted or pick up dirt. They should not be kept in a bench drawer or tool box, unless the goggles are in a good case.

A lens "fog" problem can be eliminated easily by use of one of the many commercial antifog preparations. In hot weather, a person can wear a sweatband to keep perspiration off his goggles. It takes a little effort to keep goggles clean, but the effort is no excuse for people to go without goggles and risk losing an eye.

Some people protest that goggles give them a headache. Goggles that do not fit may well produce headaches. Goggles should fit so that the right and left eyes of the wearer look through the center of the right and left lenses respectively. The nose bridge should rest flush on the bridge of the nose. The head strap on cup goggles should be adjusted for just enough tension to hold them secure, and the strap should be worn low on the back of the head.

On spectacle goggles, each bow should hook behind the ear close to the head, touching the ear all along, not just at one or two points. The spatula-type temple bow should, similarly, contact the head along its entire length, not just at one point. Correct fit is important.

If a person still complains of a headache, even though his goggles fit properly, the supervisor should direct him to see the doctor, since there may be a physiological cause for his headaches. Plano (uncorrected) spectacles have no effect on the normal eye, but people with minor eye defects may complain that they "hurt" or "draw."

"I forgot to put them on" is another excuse. This excuse may seem logical; we all forget at times. Such an excuse, however, if given again and again, usually indicates a weak safety program. The supervisor had better check that posters, bulletin boards, exhibits, and similar reminders are used, and used properly. Unless circumstances are unusual, "I forgot" should never be accepted as an excuse for not wearing safety equipment.

Contact lenses can be a source of difficulty and may lead to serious eye damage. Contact lenses should never be allowed in any industrial operation.

Protecting the Feet and Legs

About 230,000 disabling occupational foot injuries take place each year. This points to the need of foot protection in most industries, and the need for supervisors to see that their workers wear the proper protection.

All safety shoes have toes reinforced with a toe cap. The three

TABLE VI			
MINIMUM REQUIREMENTS FOR SAFETY-TOE SHOES			
Classification	Compression Pounds	Impact Foot Pounds	Clearance Inches
75	2,500	75	16/32
50	1,750	50	16/32
30	1,000	30	16/32

Source: American National Standard Z41.1–1967, *Men's Safety-Toe Footwear*.

classifications are shown in Table VI. The toe box will actually stand up under weights greater than those specified. In one case, a 10-ton roller passed over a man's toes. The toes were not crushed, but the toe box broke.

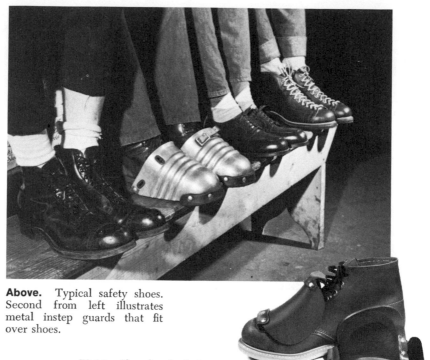

Above. Typical safety shoes. Second from left illustrates metal instep guards that fit over shoes.

Right. Shoe has built-in instep protection.

Many shoe manufacturers and jobbers cooperate with companies to help set up shoe sales departments and provide trained people to see that shoes are properly chosen for the hazard involved and are properly fitted to the wearer. Some companies keep shoes on hand for sale; others conduct business by mail. An employer can arrange for purchase of shoes at local outlets. Some dealers have a mobile shoe service—a truck equipped as a shoe store and manned by an experienced fitter.

The responsibility for proper care of safety shoes ordinarily rests with the employee.

Metal-free shoes, boots, and other footwear are available for use where there are severe electrical hazards or fire and explosion hazards.

"Congress" or gaiter-type shoes are used to protect people from splashes of molten metal or from welding sparks. This type can be removed quickly to avoid serious burns. These shoes have no laces

166

Heavy-duty leggings protect against blows and moisture. Photo was posed to show how these leggings could protect against an unsafe practice.

or eyelets to catch molten metal.

Reinforced soles or inner soles of flexible metal are used in shoes designed for use where there are hazards from protruding nails and where the likelihood of contact with energized electrical equipment is remote, for example in the construction industry.

For wet work conditions, like those found in dairies and breweries, leather shoes with wood soles, or wood-soled sandals are effective. Wood soles provide good protection on jobs that require walking on hot surfaces (that are not hot enough to char wood). Wood soles

167

have been so generally used by men handling hot asphalt that they are sometimes called "pavers' sandals."

Safety shoes with metatarsal guards should always be worn during operations where heavy materials, such as pig iron, heavy castings, and timbers, are handled. They are recommended to be worn any time there is a possibility of objects falling and striking the foot above the toe cap. Metal foot guards are long enough to protect the foot back to the ankle and may be made of heavy gage or corrugated sheet metal. They should be easy to adjust and remove.

Safety starts from top management and "works down." Here the commander of a Navy shipyard begins his tour of duty by buying safety shoes.

Overcoming objections. Safety shoes used to be hot and heavy, and people complained they were uncomfortable. New designs now make some safety shoes as comfortable, as practical, and as attractive as ordinary street shoes. The steel cap weighs about as much as a pair of rimless eyeglasses or a wrist watch. The toe box is insulated with felt to keep the feet from getting too hot or cold. Some shoes have inner soles of foam latex, which consist of tiny "breathing cells."

Some people object to wearing safety shoes because they do not cover the littlest toes. Studies, however, show that 75 percent of all toe fractures happen to the first and second toes. In most accidents, the toe box takes the load of the impact for the entire front part of the foot.

Another objection commonly voiced is that if the toe box were crushed, the steel edge would cut off the toes. Actually, accidents of this type are rare. In the majority of cases, safety shoes give sufficient protection. Certainly, freak accidents may happen, against which there is no sure protection. But, in any event, a blow that would crush the toe cap would certainly smash someone's toes if he were not wearing foot protection.

Leg protection

Leggings, which encircle the leg from ankle to knee and have a flap at the bottom to protect the instep, protect the entire leg. The front part may be reinforced to give impact protection. Such guards are used by persons who work around molten metal. Leggings should permit rapid removal in case of emergency.

Hard fiber or metal guards are available to protect shins against impact.

Knee pads protect employees whose work, like cement finishing or tile setting, requires much kneeling.

Fabrication

Foot and leg protectors are available in many different materials, the type selected depends on the work being done.

Where molten metals, sparks, and heat are the major hazards, for example, asbestos or leather is best. Where acids, alkalis, and hot water are encountered, natural or synthetic rubber or plastic, resistant to the specific exposure, may be used.

Protecting the Head

Safety hats are needed on jobs where a person's head is menaced by falling or flying objects or by bumps. American National Standard

Z89.1–1969, *Safety Requirements for Industrial Head Protection,* already mentioned, gives specifications that a protective hat must meet.
Impact resistance is essential. Hats must be fire resistant and impervious to moisture. Where contact with energized circuits is possible, only hats with shells made of nonconductive material should be worn. These hats should have no conductive fittings passing through the shell. Hats are available for utility service with a dielectric strength of 20,000 volts.

A brim all around the hat provides the most complete protection for the head, face, and back of the neck. In situations where the brim would get in the way, the cap type may have to be worn.

In addition to these, there is another type of head protection available. Known as "bump hats" or "bump caps," these are for use in confined spaces where the exposure is to bumping only. They should never be used on construction sites or shipyards, or other locations where there is exposure to any hazard greater than bumping the head against an obstruction.

Fabrication

Laminated plastic, molded under high pressure, is widely used for safety hats. It resists impact, water, oil, and electricity. Glass fiber impregnated with resin is widely used because of its high strength-to-weight ratio, high dielectric strength, and high resistance to moisture. Glow-in-the-dark hats are available on special order and give added protection against traffic accidents where employees work at night or in darkened areas.

Aluminum alloy hats protect against impact but not against electric shock. They should not be worn where there is danger of electrical contact. Their use is diminishing in favor of other types.

For electrical work, safety hats should be tested periodically, in accordance with American National Standard Z89.2–1971, *Safety Requirements for Industrial Protective Helmets for Electrical Workers.* They may fail this test before their mechanical strength is appreciably reduced. If they fail the test, the hats should be plainly and permanently marked for use only in mechanical work.

The hard outer shell of the hat is supported by a suspension—a cradle attached to a headband which keeps the shell away from the head and cushions any blows. This must be adjusted to fit the wearer

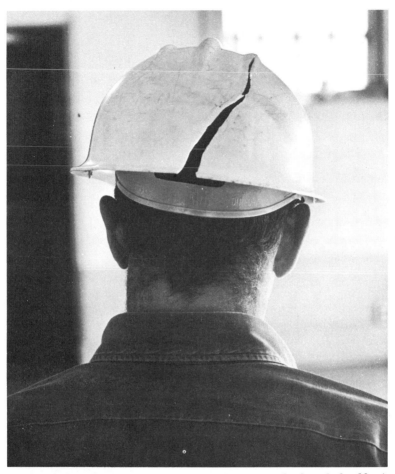

A three-day headache (and some minor bruises on back and shoulders) was all this communications lineman received when a 500-pound, 25-foot pole fell on his head. The brunt of the blow was caught on this company-issued safety hat. The impact pushed the lineman's head against the ground and split the hat (as shown). The pole came to rest on the lineman's left shoulder.

well. Cradles and headbands should be kept on hand and should be replaced whenever they show signs of deterioration or bad soiling.

If bands and cradles are not replaced, the band, cradle, and shell should be washed in warm, soapy water and then rinsed. Steam cleaning is also effective. Thirty days is the recommended maximum interval between cleanings. Hats and suspensions should be thoroughly cleaned before being reissued.

No attempt should be made to repair the shell of a hat once it has been broken or punctured. A few extra shells should be kept on hand as replacements. Do not let people drill holes in their safety hat to "improve ventilation" or let them cut notches in the brim. Such practices destroy the ability of the hat to protect the wearer.

Auxiliary features

Liners are available for cold weather use. Do not let workmen remove the safety hat suspension so they can wear the hat over a parka hood. This practice completely destroys the protection given by the hat, and has led to tragic results.

A chin strap is useful when the wearer may be exposed to strong winds, such as on bridges or oil derricks.

An eye shield of transparent plastic can be attached to some types of hats. The shield is hinged under the brim and lies flat against the brim when not in use.

Color. White, grey, red, green, blue, brown, and black are the standard hat colors. Other colors are available on special order. Colors that go all the way through the laminated shell are permanent. Painting permits a greater selection of color and design, but is not as permanent. Often distinctive colors or designs are used to designate the wearer's department or trade, especially in companies or plants where certain areas are restricted to a few selected employees.

Brackets to support welding masks or miners' cap lamps are available on some safety hats.

Overcoming objections

A common complaint against wearing safety hats is that they are too hot in the summer. Actually, the hat is cooler than a tight-fitting cap or felt hat that many men wear. Ventilation, provided by the air spaces between the cradle and the shell, keeps the head cool and comfortable. The hat surface, too, reflects heat.

In winter, the complaint is that the hats are too cold. A liner or skull cap large enough to come down over the ears may be worn. The suspension, as mentioned, must not be removed.

Using a sound level meter, a safety professional measures noise level at various work areas to determine whether the exposure is great enough to require personal protection.

Some people complain that a hat gives them headaches. It is possible that it may, until the wearer becomes accustomed to wearing it. There is no physiological reason, however, why a properly fitted hat should cause a headache.

Safety hats are slightly heavier than ordinary hats, but this difference is not noticed after a while if the headband and suspension are properly adjusted, unless, of course, there may be other causes—physiological or psychological.

Ear Protection

Excessive noise should be reduced whenever possible, as outlined in the previous chapter. Ear protection should be used only as a last resort.

Although there is some disagreement as to the maximum intensity of sound to which the human ear can be subjected without damage to hearing, the standards of the Occupational Safety and Health Act should be used. The permissible noise exposure for an 8 hour duration is 90 dB, measured on a weighted scale that corresponds to the response of the human ear. This level may be increased slightly as the duration of exposure decreases. (Review the discussion in the previous chapter, pages 121 to 123.)

A single decibel (dB) represents the smallest change in volume of sound that can be detected by the human ear. A 130 dB sound is so intense that it produces a sensation of pain in the ear. The decibel is a ratio-type scale. If the intensity of a sound is doubled, it increases 3 dB; if halved, it decreases 3 dB. If cut to one-tenth, it decreases by 10 dB; if cut to one-one hundredth, it decreases by 20 dB. The human ear is not as sensitive as a sound-measuring device; it interprets a 10 dB drop in intensity as being "half as loud."

Every effort should be made to reduce the sound level by engineering control, but where this is not possible, personal protective equipment shall be provided.

It takes specialized equipment and trained men to analyze a noise exposure. See the photo on the previous page. It takes a physician or company medical department to approve types of ear protection that will be most effective.

Amount of protection

Commercially available earplugs, if properly fitted and used, generally reduce noise reaching the ear by 25 to 30 dB in the higher (more harmful) frequencies. They will give ample protection against sound levels of 115 to 120 dB. Earmuffs of the better type can reduce noise an additional 10 to 15 dB, making them effective against sound levels of 130 to 135 dB.

Combinations of earplugs and muffs give 3 to 5 dB more protection. In no case, however, will total attenuation (sound reduction) be more than about 50 dB, because conduction of noise through the bones of the head becomes significant at this point.

The best ear protection is the one that is accepted by the individual and is worn properly. Properly fitted protectors can be worn continuously by most persons and will provide adequate protection against most industrial noise exposures.

Insert-type protectors

This type is inserted into the ear canals and varies considerably both in design and material. Materials used are pliable rubber, soft or medium plastic, wax, and cotton. (*a*) Rubber and plastic types are popular because they are easy to keep clean, are inexpensive, and give good performance. (*b*) Wax tends to lose its effectiveness during the work day because jaw movement changes the shape of the ear canal and this breaks the acoustic seal between ear and insert. Wax inserts may be objectionable for use in dirty areas because they must be shaped by hand. They should be used only once. (*c*) Cotton is a poor choice because of its low attenuating properties and because it must be hand formed. Glass down, so-called "Swedish cotton," is being used with success by some.

Still another type that seems to be gaining in popularity is one that is specially molded to each ear. After being allowed to set, it will hold its form. Because each person's ear canal is different, these plugs become the property of the individual to whom they were fitted. These plugs must be fitted by a trained, qualified person.

Because of the pressure required to fit a rubber or plastic earplug into the ear canal so that the noise does not leak around the edges (as much protection as 15 dB could be lost), points of pressure develop, which may cause discomfort. To overcome this problem, plugs should be fitted individually for each ear. A good seal cannot be obtained, however, without some initial discomfort. There will be no lasting reactions as a result of using earplugs if they are made of soft material and kept reasonably clean. Hard, rigid materials could injure the ear canal.

Skin irritations, injured ear drums, or other harmful reactions are exceedingly rare when properly designed, well fitted, and clean ear protectors are used. They should cause no more difficulty than does a pair of well fitted safety goggles.

If fitting continues to be a problem, casts can be made of the ear canals and a plastic plug can be custom made for each.

Overcoming objections. Although some people object to earplugs because they believe they will not be able to understand conversation, tests show that when noise level is higher than about 85 dB, speech is more easily understood with earplugs in place than it is without them. At worst, a person wearing earplugs is like one who is moderately hard of hearing, rather than deaf.

175

Supervisors should be aware of noise hazards in their departments. They should also be able to show the correct use of earplugs when other forms of control are not practical.

Muff types

Cup or muff devices cover the external ear to provide an acoustic barrier, the effectiveness of which varies with the size, shape, seal material, shell mass, and type of suspension of the muff. Head size and shape also influence the effectiveness. Liquid- or grease-filled cushions give better noise suppression than plastic or foam rubber types. Muffs are made in a universal type or a head, neck, or chin type.

Protecting Fingers, Palms, and Hands

Fingers and hands are exposed to cuts, scratches, bruises, and burns. Although fingers are hard to protect (because they are needed for practically all work), they can be shielded from many common injuries by use of proper protective equipment, such as the following.

1. Asbestos gloves protect against burns and discomfort when the hands are exposed to sustained conductive heat.
2. Metal mesh gloves, used by those who work constantly with knives, protect against cuts and blows from sharp or rough objects and tools.
3. Rubber gloves are worn by electricians. They should be tested regularly for dielectric strength.
4. Rubber, neoprene, and vinyl gloves are used when handling chemicals and corrosives. Neoprene and vinyl are particularly useful when petroleum products are handled.
5. Leather gloves resist sparks, moderate heat, chips, and rough objects. They provide some cushioning against blows. They are generally used for heavy-duty work. Chrome-tanned leather or horsehide gloves are usually used by welders.
6. Chrome-tanned cowhide leather gloves with a steel-stapled leather patch or steel staples, on palms and fingers, are used in foundries and steel mills.

Full-hand metal-mesh glove and abdominal protector reduces knife injuries. Supervisors must make sure equipment is always in good condition; slight defects or worn equipment can become a real hazard.

7. Cotton or fabric gloves are suitable for protection against dirt, slivers, chafing, or abrasion. They are not heavy enough for use with very rough, sharp, or heavy materials.
8. Coated fabric gloves protect against moderately concentrated chemicals. They are recommended for use in canneries, packing houses, food handling, and similar industries.

Specially made, electrically tested rubber gloves, worn under leather gloves which prevent punctures, are used by linemen and electricians who work with high voltage equipment. A daily visual inspection and air test (by mouth) must be made. Make sure that rubber gloves extend well above the wrist so that there is no gap between the coat or shirt sleeve and glove. The shortest glove is 14 in.*

People usually do not mind wearing gloves when the dangers of not wearing them are apparent. When a worker no longer has to worry about cutting his hands, and when he can grip materials better because of his gloves, production increases. Still, if employees become lax in wearing their gloves, supervision is needed.

The supervisor should be aware that gloves may be a hazard when worn around certain machining operations. Their use here should not be allowed.

Hand leathers or hand pads are often more satisfactory than gloves for protecting against heat, abrasion, and splinters. Wristlets or arm protectors are available in the same materials as gloves.

Protective creams may be used to protect skin against many irritants when protective equipment is not practicable. Creams are available in water-soluble or water-resistant types, and in several grades for different exposures.

Water-soluble creams protect against cutting oils, paints, lacquers, and varnishes. Water-resistant types are used where the cutting oil or cooling lubricant has more than 10 percent water content. Soap and warm water will remove creams of this type.

For protective cream to be effective, coatings should be renewed frequently. It should be remembered that creams do not protect against highly corrosive substances.

Protecting the Torso

The most common protection for the abdomen and trunk is the full apron. Aprons are made of various materials. Leather or fabric aprons, with padding or stays, offer protection against light impact and against sharp knives and cleavers, such as used in packing houses. Asbestos coats and aprons are often used by those who work around hot metal or other sources of intense conductive heat.

An apron worn near moving machinery should fit snugly around

* See NSC Industrial Data Sheet No. 598, "Flexible Insulating Protective Equipment for Electrical Workers."

the waist. Neck and waist straps either should be light strings or instant-release fasteners in case the garment is caught. There should be a fastener at each end of the strap to prevent severe friction burns should the strap be caught or drawn across the back of the neck. Split aprons should be worn on jobs that require mobility on the part of the worker. Fasteners draw each section snugly around the legs.

Welders are often required to wear leather vests or capes and sleeves, especially when doing overhead welding, as protection against hot sparks and bits of molten metal.

On jobs where men must carry heavy and angular loads, pads of cushioned leather or padded duck are often used to protect the shoulders and back from injury.

Respiratory Protective Equipment

Respiratory equipment must be regarded as emergency equipment, or equipment for occasional use. The first consideration where there are contaminants is to remove them at the source or isolate the process. Since leaks and breakdowns may occur, however, and since some operations expose a person only briefly and infrequently, respiratory equipment should be available. The user shall be instructed and trained in its proper use and its limitations.

Selecting the respirator

Air contaminants range from relatively harmless substances to toxic dusts, vapors, mists, fumes, and gases that may be extremely harmful. Where equipment may be needed, the chemical or other offending substance must be determined, and the extent of the hazard evaluated. With this information, the respirator that protects against the particular hazard can be selected. Before respiratory equipment is ordered, it is best to discuss the type of exposure with the company safety department, and with manufacturers and dealers.

Equipment should bear the National Institute of Occupational Safety and Health approval; equipment used in mines carries United States Bureau of Mines approval. This assures the design, durability, and workmanship of equipment. Worker safety, freedom of movement, vision, fit and comfort of facepiece and headpiece, the ease with

179

FILTER
RESPIRATOR

GAS
MASK

AIR LINE
RESPIRATOR

*1. Compressed
oxygen (cylinder)
rebreathing
apparatus*

*2. Compressed oxygen
or air (cylinder)
nonrebreathing
apparatus*

*3. Oxygen-generating
(rebreathing)
apparatus*

SELF-CONTAINED   BREATHING   APPARATUS

HOSE
MASK
Supplied by blower

Several types of respiratory protective equipment. Upper left: twin-cartridge filter respirator. Upper center: canister gas mask. Upper right: air line respirator. Center left: compressed oxygen (cylinder) rebreathing apparatus. Middle center: compressed oxygen or air (cylinder) nonrebreathing apparatus. Right center: oxygen generating (rebreathing) apparatus. The units in the middle row are self-contained. At the bottom is a hose mask, supplied by blower.

which the filter and other parts can be replaced, the dust-tightness of the apparatus, freedom from leakage, and resistance to flow of air when the wearer is inhaling and exhaling are all tested.

Types of equipment

Respiratory equipment includes mechanical filter respirators, chemical cartridge respirators, gas masks with canisters, hose mask and air line respirators, and self-contained breathing apparatus.

Mechanical filter respirators must protect against exposure to nuisance dusts and pneumoconiosis-producing dusts, mists, and fumes. Examples of nuisance dusts are aluminum, cellulose, cement, coal, flour, gypsum, iron ore, limestone, and sawdust.

Pneumoconiosis comes from three Greek words that mean "lung," "dust," and "abnormal condition." The generally accepted meaning of the word is merely "dusty lung." The kind of dust inhaled determines the type of condition or injury. A number of organic dusts are capable of producing lung diseases, but not all these diseases are classified as pneumoconioses because they are not all a "dusty condition" of the lung.

In very rare cases, enough dust has been inhaled to cause mechanical blockage of the air spaces. Flour dust has been known to cause this condition. Some dusts may be essentially inert and remain in the lungs indefinitely with no recognizable irritation, and a few (like limestone dust) may gradually dissolve and be eliminated without harm.

Another type of filter respirator is approved for toxic dusts, such as lead, asbestos, arsenic, cadmium, manganese, selenium, and their compounds.

Protection against mists, such as chromic acid, and exposure to fumes, such as zinc and lead, is also afforded by types of mechanical filter respirators specifically approved for such exposures. The filter, usually made of paper or felt, should be replaced or cleaned frequently, for if it becomes clogged, it restricts breathing or becomes inoperative. This may happen as frequently as several times a shift.

A mechanical filter respirator is of no value as protection against solvent vapors, injurious gases, or oxygen deficiency. To use it under these conditions is a common, but most serious mistake.

Chemical cartridge respirators usually have a half-mask facepiece

181

Chemical cartridge-type mask protects against insecticide and pesticide. For complete protection, man should wear a head covering.

(or a full-mask facepiece) connected to one or more small containers of activated charcoal or soda lime (a mixture of calcium hydroxide with sodium or potassium hydroxide) for absorption of low concentrations of certain vapors and gases. These concentrations are, approximately, 0.10 percent (by volume) for organic vapors, 0.05 percent for acid gases, 0.05 percent for combinations of organic vapors and acid gases, and 0.07 percent for ammonia. The life of the containers is very short. For protection against mercury vapors, the nominal container life is 8 hours. After use, the containers must be discarded.

These respirators must not be used in atmospheres immediately dangerous to life, such as those deficient in oxygen.

A gas mask consists of a facepiece or mouthpiece connected by a flexible tube to a canister. Inhaled air, drawn through the canister, is cleaned chemically. No one chemical has been found that removes all contaminants, unfortunately. Therefore, the canister must be chosen to meet the exposure.

Gas mask canisters are identified by their color as to type of exposure.*

Black—organic vapors
White—acid gases
Yellow—organic vapors and acid gases
Green—ammonia gas
Brown—organic vapors, acid gases, and ammonia gases
Red with grey stripe near top—universal. All industrial gases (including carbon monoxide, smoke, and fumes)
White with green stripe near bottom—hydrocyanic acid (hydrogen cyanide) vapor
White with yellow stripe near bottom—chlorine gas
Blue—carbon monoxide
Purple—radioactive materials, except tritium and noble gases.

Gas masks have definite limitations on their effectiveness, both gas concentration and length of time influence this. Gas masks, like chemical cartridge respirators, do not protect against oxygen deficiency.

When a canister is used up, it should not be left attached, but should be removed and replaced by a fresh one. Even if they have not been used, canisters should be replaced periodically. Manufacturers will indicate the maximum effective life of their canisters.

Gas masks must be quickly available for emergencies. For example, in an ice plant where an ammonia leak is likely, masks should be placed either just inside or just outside the exit doors so that they can be reached quickly without a person being first disabled by the gas. Masks should be stored away from moisture, heat, and direct sunlight. They should be inspected regularly.

Hose masks are used by men entering tanks or pits where there may be dangerous concentrations of dust, mist, vapor, or gas, or insufficient oxygen. No one should enter, much less remain in a tank (or similar space) that tests show has less than 16 percent oxygen at

* American National Standard K13.1–1967, *Identification of Gas-Mask Canisters.*

any time, unless he wears approved respiratory protective equipment, such as a hose mask or self-contained breathing apparatus. The atmosphere of the enclosed space should be tested to see if there are any toxic or flammable contaminants present in dangerous or explosive concentrations. If so, or if oxygen is deficient, the space should be ventilated before a man enters. The enclosure should be retested as long as men must work in it, and reventilated if necessary.

The hose mask consists of a facepiece with head harness, valves for inhaling and exhaling, and one or more breathing tubes. This tube connects to a hose supported by a body harness. The hose reaches to an intake where a supply of fresh air is available.

In the absence of a blower, not more than 25 ft of hose should be used. With a blower, hose masks are approved for use with up to 300 ft of hose, but in no case under conditions that make the exhalation resistance over 1 in. of water, or the inhalation resistance over 4 in. of water. The blower must be operable by hand in case of power failure.

The hose mask is used where there is a known exposure to harmful material in the atmosphere or where there may be insufficient air to support life. A hose mask with blower or self-contained breathing apparatus must be used where, in the event of failure of the equipment, hasty escape from the dangerous atmosphere would be impossible or would result in serious injury.

With either type, the user must wear a safety harness with life line attached and must be constantly attended by another man similarly equipped. In case of failure of the air supply in a hose mask with a blower, the wearer can still breathe through the hose while making his escape.

The air line respirator can be used in atmospheres not immediately dangerous to life, especially where working conditions demand continuous use of a respirator. Each person should be assigned his own respirator.

Air line respirators are connected to a compressed air line. A trap and filter should be installed in the compressed air line ahead of the masks to separate oil, water, grit, scale, or other matter from the air stream. When line pressures are over 25 psi (gage), a pressure regulator is required. A pressure release valve, set to operate if the regulator fails, should be installed.

To get clean air, keep the compressor intake away from any source of contamination, such as internal combustion engine exhaust. The

Oxygen-generating (recirculating) apparatus, one-hour type.

Compressed air (cylinder) nonrebreathing apparatus.

compressor should have a temperature regulator, or the compressed air line should have a carbon monoxide alarm in order to guard against the carbon monoxide hazard from overheated lubricating oil or from engine exhaust. The most desirable air supply is provided by a nonlubricated or externally lubricated medium-pressure blower, such as a rotary compressor.

If a man must move from place to place, he may find the air hose a nuisance. The supervisor must realize that this will reduce the man's efficiency. Care must be exercised to prevent damage to the hose; for example, it should not be permitted to lie in oil.

The abrasive blasting helmet is a variety of air line respirator designed to protect the head, neck, and eyes against the impact of the abrasive, and to give a supply of breathing air. The respiratory requirements are the same as those described for air line respirators.

The helmet should be covered both inside and out with a tough, resilient material. This increases comfort and still resists the abrasive. Some helmets have an outer hood of impregnated material and a zippered inner cape for quick removal. A glass window, protected by a 30- to 60-mesh fine-wire screen, should be provided. Safety glass,

185

used to prevent shattering under a heavy blow, should be free of color and glass defects.

Self-contained breathing apparatus. When a man must work in a hazardous atmosphere at distances greater than 300 ft from a source

Rescue and emergency use of self-contained breathing apparatus is dramatized by contest to see who in this refinery can get into the 30-pound air-pack equipment the fastest. Here, the competitive champ shows how he does it in 16 seconds flat. With practice, most employees could rig up in less than one minute.

of fresh air, he should wear a self-contained breathing apparatus, which carries its own supply of oxygen. This equipment is frequently used in mine rescue work and in fire fighting.

The three principal types of self-contained breathing apparatus are the compressed oxygen rebreathing type, the compressed oxygen (or air) nonbreathing type, and the oxygen-generating type. The length of time these units may be used is limited strictly by NIOSH and Bureau of Mines specifications.

Self-contained breathing apparatus should be worn only by men who are physically fit and well trained. Men should be refresher trained at least every six months in order to maintain efficiency.

Because of the extreme hazard, no one wearing self-contained breathing apparatus should work in an irrespirable atmosphere unless other persons similarly equipped are standing by, ready to give help.

Cleaning the respirator

Respirator facepieces and harnesses should be cleaned and inspected regularly. If several persons must use the same respirator, it should be disinfected after each use in order to comply with OSHA regulations. Methods of disinfection include:

1. Immersion in a weak solution of quarternary ammonium compound, followed by a warm-water rinse. This solution is not injurious to skin or to rubber.

2. Washing in warm, soapy water and rinsing for at least one minute in clean water of 140 F minimum temperature.

The supervisor should inspect respirators at intervals to check for damage or improperly functioning parts, such as headbands or valve seats.

Overcoming employee complaints

Many people consider respirators a nuisance because they do not realize that failure to wear them may endanger their lives. The supervisor should tell the men why wearing the equipment is necessary, show them how to fit it in position, and explain its operation.

Another objection is that the respirator is difficult to breathe through. A properly maintained, NIOSH-listed respirator has minimum breathing resistance.

If the respirator is removed at intervals, dust settles on it. When the respirator is replaced, these dust particles are transferred to the skin and cause irritation. To prevent this, the wearer should keep the respirator on constantly when in a contaminated atmosphere. To further reduce irritation, ointments or jellies, or a cotton facelet, may be used.

Safety Belts and Harnesses

Safety belts and harnesses with life lines attached should be worn by those who work at high levels or in closed spaces where the air supply may not be adequate, and by those who work where they may be buried by loose material or be injured in confined spaces. This discussion does not include seat (vehicular safety) belts or linemen's belts.

Both normal and emergency types are available. Normal use involves comparatively light stresses applied during regular work—stresses that rarely exceed the static weight of the user. Emergency use means stopping a man when he falls—every part of the belt may be subjected to an impact loading many times the weight of the wearer.

A window cleaner's belt, for example, is subjected to a moderate load most of the time it is used. It will be subjected to a severe loading, however, if the man falls when only one terminal of the belt is attached. A belt for a person who leans back as he works should, therefore, have two D-rings, one on each side of the belt, to which a throw rope or lanyard can be attached. The rope then is connected to an anchorage.

A **boatswain's swing chair** is used where the entire weight of a man must be supported vertically, as in raising and lowering him along a wall. The chair has one strap (sometimes with a board attached) to sit on, and a waist strap fastened on each side of the man so he can either stand or sit, but cannot fall out of the seat. The chair and fittings must conform to OSH regulations §1910.28(j).

A **harness-type safety belt** better distributes the shock of an arrested fall. The shock is distributed over shoulders, back, and waist, instead of being concentrated at the waist. The harness permits a man to be lifted with his back straight, rather than bent over a waist strap. This makes rescue easier if the victim is unconscious, buried, or must be taken out through a manhole. Wherever a job requires use of a hose mask, a harness and lifeline should also be used.

If long free falls are possible, the harness should be designed to distribute the impact force over the legs and chest as well as the waist. A shock absorber or decelerating device, which brings the falling man to a gradual stop, lessens the impact load on both the equipment and the man. To prevent a long fall, the line should be tied off overhead

188

Safety belts and harnesses with life lines attached should be worn by all who work at high levels, or in tanks or other confined areas. Be sure to inspect belts before use.

and should be as short as movements of the worker will permit.

Fabrication

Both leather and webbing belts are furnished by most manufacturers. Nylon belts, straps, and harnesses are also available, but are not satisfactory under all conditions. The manufacturers should be consulted before such equipment is ordered.

In general, webbing is superior to leather for any safety belt that may have to absorb impact loads. Webbing has three to four times as much resistance to impact loading as leather of the same size, because webbing has both more strength and more stretch. Webbing belts normally use friction buckles, which avoid loss of strength at buckle holes. Leather requires special care and treatment to retain its strength; webbing does not.

Commercial grade harness leather will be cut through by a single-tongue buckle at a loading of 300 to 500 lb; this is adequate for lifting a man from a tank or bin. In contrast, web belts can be obtained that possess strengths of over 12,000 lb.

Although well tanned and well oiled leather is not easily attacked by most chemicals, it should not be left in contact with them. Regular cleaning after use is important. Webbing will stand more heat than leather, and when soaked in water, will dry to its natural condition.

In order to check a fall, belts demand strength proportionate to the possible distance of the fall and the weight of the falling body. A 2-in. by ¼-in. leather would probably arrest the fall of a window washer to 6 ft. It might break at 7 ft. Therefore, the belt should be at least 3 in. by ¼ in., with a double-tongue buckle. A herringbone-web belt affords a much greater safety margin.

Special types of webbing are available for certain uses. Wax-treated webbing resists paint and mildew. Neoprene-impregnated webbing resists acid and oily conditions, a property important to the chemical and petroleum industries.

The most common types of webbing are the square or basket weave and the herringbone weave, which has about twice the strength of the basket weave of the same size and thickness.

Life lines

Manila rope of ¾-in. diameter, or nylon rope of ½-in. diameter is recommended for life lines. Nylon is more resistant to wear or abrasion than is manila (abaca), it is more elastic (absorbing shocks and sudden loads better), and has high tensile strength (wet or dry). Nylon is tough, flexible, and easy to handle. Because it resists mildew, it can be stored wet.

Breaking strength is always measured in a straight-line pull on the rope. Half-inch diameter manila rope breaks at approximately 2,650 lb; half-inch diameter nylon rope is rated at 6,400 lb (± 5 percent). Using a safety factor of 5 for manila and 9 for nylon gives safe load strengths of 530 lb and 710 lb, respectively.

Manila rope ¾-in. in diameter breaks at approximately 5,400 lb, giving a safe load strength of 1,080 lb. When a lowload-limit shock absorber is used, ¾-in. manila rope or ½-in. nylon rope is considered strong enough for most life lines. Without a shock absorber, even ¾-in. manila may not be strong enough to arrest a long fall.

Wire rope should not be used as life lines where free falls are possible, unless a shock-absorbing device is also used. The rigidity of

TABLE VII	
APPROXIMATE EFFICIENCY OF MANILA ROPE HITCHES	
	Percent
Full strength of dry rope	100
Eye splice over metal thimble	90
Short splice in rope	80
Timber hitch, round turn, half hitch	65
Bowline, slip knot, clove hitch	60
Square knot, weaver's knot, sheet bend	50
Flemish eye, overhand knot	45
Source: Oregon Safety Code for Places of Employment.	

steel greatly increases the impact loading. Steel is also hazardous around electricity.

Knots reduce the strength of all ropes. The degree of loss depends upon the type of knot and the amount of moisture in the rope. See Table VII. More information on pages 258 to 262.

Inspection and care of belts and lines

Belts and lines should be looked over each time they are to be used. At least once every three months, belts should be examined by a trained inspector.

Belts. Leather belts must be carefully checked for cuts and

191

scratches on either side. Any deep vertical cut (crosswise to the belt) justifies discard. Lengthwise cuts are not as serious, but should be viewed critically, and unless they are small, the belt should be discarded. Web belts should be examined for worn and torn fibers. When a number of the outer fibers are worn or cut through, the belt should be discarded.

Belt hardware should be examined and worn parts replaced. If the belt is riveted, each rivet should be carefully inspected for wear around the rivet hole.

Dirt and dust should be brushed carefully from leather belts to prevent scratching. The belt should then be washed with warm water and castile or saddle soap, then rinsed in clear warm water, and dried at room temperature. Before it is completely dry, the belt should be oiled with neatsfoot oil, castor oil, soybean oil, or special silicone compound, but never mineral oil.

Web belts may be washed in warm, soapy water, rinsed with clear water, and dried by moderate heat. If these belts are worn under unusual conditions, or if a dressing is to be used, consult the manufacturer.

Lines. The outer surface of rope lines should be examined for cuts and for worn or broken fibers. Manila rope should be discarded when it has become smaller in diameter or has acquired a smooth look. Inner fibers of manila rope should be examined for breaks, discoloration, and deterioration. If it shows any of these signs, the rope should be discarded.

A steel wire rope should be examined for broken strands, rust, and kinks that may weaken it. Ropes must be kept clean, dry, and rust free. They should be lubricated frequently, especially before use in acid atmospheres or before exposure to salt water. After such use, wire rope should be carefully cleaned and again coated with oil.

Protection Against Ionizing Radiation

Ionizing radiation, dangerous because of its serious biological effects, need not be feared if it is respected and if suitable precautionary measures are taken. NSC's *Accident Prevention Manual* and *Fundamentals of Industrial Hygiene* have much information and many references.

Coping with radiation hazards requires special clothing and equipment. As part of his training course, this student radiological monitor enters a simulated laboratory to make instrument readings. Smoke and radioactive source materials add to the realism of this exercise.

Standards for protection against radiation, specified in OSH regulations §§1910.96 and 1926.53, which reference Atomic Energy Commission regulations spelled out in *Code of Federal Regulations*, Title 10, Part 20, should be followed.

Rubber gloves for handling radioactive materials, disposable clothing, suits with supplied air, and approved respiratory devices are some of the items of specialized equipment that are needed. Under no circumstances should contaminated clothing be worn into clean areas.

Thorough washing with soap and water is the best general method for decontamination of the hands and other parts of the body, regardless of the contaminant.

Before any supervisor or employee works with radioactive materials, he should check with his medical and safety departments. He should be assured in writing that all people will be adequately protected before any tests are made or procedures changed that involve radioactive materials or isotopes. Supervisors should see that medical clearances and other authorizations, in addition to badges and dosimeters, are issued to all personnel assigned to work in areas restricted for radioisotope use.

Monitoring radiation

Employees should be continuously monitored for exposure to radiation. At work, they should wear film badges, which should be developed and replaced at regular intervals, depending on the radiation intensity level to which they are exposed.

The film badge records the total exposure for a given period, but does not indicate the precise time during the period when the exposure occurred.

For better protection, each person should also wear two pocket dosimeters. These show the dosage a person has received during any part of a work period so that he can check at any time for exposure. The dosimeter does not provide a permanent record, but is essential for safety during work periods. It is read at least once, usually at the end of a man's shift. It can be reset by means of a dosimeter charger.

Radiation accidents resulting in overexposures should be immediately reported to the medical director.

Safe Work Clothing

Ordinary work clothing, if clean, in good repair, and suited to the job, may be considered safe. "Safety clothing" refers to garments designed for specific, hazardous jobs where ordinary work clothes do not give enough protection against minor injuries, such as abrasions, burns, and scratches.

Good fit is important. Most work pants are made extra-long to provide for shrinkage and to fit tall men. Trousers that are too long must be shortened to proper length, preferably without cuffs. If cuffs are made, they should be securely stitched down so the wearer cannot catch his heel in them. Cuffs should never be worn near operations that produce flying embers, sparks, or other harmful matter that might get caught in them.

Neckties, long or loose sleeves, gloves, and loose-fitting garments (especially about the waist) create a hazard, because they are easily caught in moving (or revolving) machine parts.

All types of jewelry are out of place in a shop—rings, bracelets, and wrist watches can cause serious injury. A finger may be torn off if a ring catches on a moving machine part, or on a fixed object when the body is moving rapidly. Necklaces, key chains, and watch chains also constitute hazards near moving machinery. Metal jewelry worn around electrical equipment, including batteries, can be dangerous.

Clothing soaked in oil or flammable solvent is easily ignited and thus a definite hazard. Skin irritations are often caused by continued contact with clothing that has been soaked with solvents or oils.

If oil or dust may get in the hair, a cap should be worn to guard against dirt and infections and to keep hair looking nicer. A cap may also help protect a person from moving parts of machines that cannot be completely guarded. (More about this later.)

Materials for protective clothing

A number of protective materials are used in making clothing to protect workers against various hazards. Supervisors should be familiar with these materials.

Aluminum-faced fabric has a coating that reflects radiant heat. Aluminized asbestos is used for heavy-duty suits, and aluminized fabric for fire-approach suits.

Asbestos is used as protection against intense conducted heat and against flames. When used with a radiation barrier of reflective material, asbestos offers excellent protection in fire-fighting and rescue work.

Flame-resistant cotton fabric is often used as a hair covering for

195

people who work near sparks and open flames. Although the fabric is durable, the flame-proofing treatment may have to be repeated after from one to four launderings.

Flame-resistant duck, used for garments worn around sparks and open flames, is lightweight, strong, and long-wearing. It is not considered adequate protection against extreme heat.

Glass fiber is used in multi-layer construction to insulate clothing. The facing is made of glass cloth or of aluminized fabric.

Impervious materials (such as rubber, neoprene, vinyl, and fabrics coated with these materials) protect against dust, vapors, mists, moisture, and corrosives. Rubber is used widely because it resists acids, caustics, and other corrosives. Neoprene resists petroleum oils, solvents, acids, alkalis, and other corrosives.

Leather protects against light impact. Chrome-tanned leather protects against sparks, molten metal splashes, and infrared and ultraviolet radiation.

Synthetic fibers (such as Orlon, Dynel, and Vinyon) resist acids, many solvents, mildew, abrasion and tearing, and repeated launderings. Because some fabrics generate static electricity, garments should not be worn in explosive or high-oxygen atmospheres unless they are treated with an antistatic agent.

Water-resistant duck is useful for exposures to water and noncorrosive liquids. When aluminum coated, the material also protects against radiant heat.

Wool may be used for clothing that protects against splashes of molten metal and small quantities of acid, and small flames.

Permanent press. With the wide use of the permanent press garments, care should be taken because some of this clothing is quite flammable.

Clothing for women

Women need special types of head guards to keep their hair from catching in lathes, drill presses, sewing machines, and other kinds of moving machinery, especially that which produces enough static electricity to raise the hair. Caps also protect hair against oil and dust.

Caps must enclose every wisp of hair, and have a visor long enough to provide a warning if the head comes close to machinery.

Hair nets, scarves, and babushkas are satisfactory head protection on many jobs other than those around moving machinery.

Headgear worn around sparks or flames should be made of flame-resistant material.

If women work in areas where men are required to wear safety hats or caps, women too must wear the same protection.

Once women are convinced that protective headgear helps keep

Cap affords hair protection for women in general factory work. Generous visor on cap serves as a feeler guard to prevent head injuries resulting from contact with objects.

their hair clean, attractive, and free from dust and oil, the problem of getting them to wear protective hair covering becomes easier.

Slacks are required to be worn by women who work around machinery.

Women's work shoes should be sturdy, comfortable, and well fitting. Heels should not be more than 1¼-in. high. Open-toe shoes should not be worn by women in industrial operations. Safety shoes should be worn where hazards so indicate.

Long-haired (hirsute) men have similar problems. See how to handle them next.

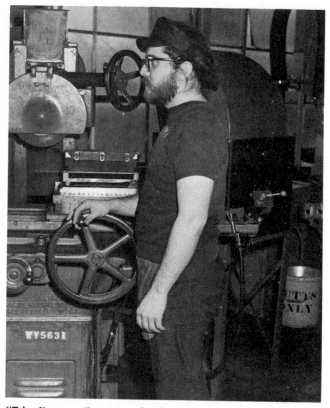

"Trim it or confine it" is this firm's safety policy. Beard bags are also required for long beards. Fortunately, injuries attributed to long hair, sideburns, moustaches, and beards are not too common. When they do occur, however, they are usually very painful and generally receive a lot of publicity.

Special problems today for men

At one time it was felt that only women were more concerned about their appearance than men, so their work clothing should be attractive and well fitting; today this may be a fallacy because men seem to be as clothes-conscious as women.

Many men also wear their hair quite long. These individuals present a hazard to themselves when working under certain conditions. If they are working on any machine with revolving, rotating, or spinning parts, or in an operation where their hair might become entangled, they should be required to wear the same type protection that women must wear on the same operation—a hair net, safety cap, snood, or similar protection.

Men with extremely long and heavy sideburns sometimes find it impossible to get a proper fit when wearing ear muffs. If a high degree of noise attenuation is required, it may be necessary to have the employee cut his sideburns.

Large bushy mustaches and beards also can be a hazard. It has been proven that it is impossible to get a proper fit with many types of face pieces used with respiratory equipment and breathing apparatus. Here again certain rules must be made and adopted and employees required to conform to these rules.

Because a person's life may depend upon the piece of respiratory equipment he is wearing, it is vitally important that it fit properly.

Some types of "Afro" hair styles make it impossible for a person to properly fit a safety hat to his head with the result that he either can't wear it at all or he must remove the suspension cradle which, of course, defeats the sole purpose of wearing the hat.

Industry must establish some guidelines for proper dress. Employees must understand the need for these rules and standards, and they must agree to follow them.

'Selling' Personal Protective Equipment

To help sell the use of personal protective equipment, the supervisor should be familiar with various techniques used successfully by other companies. These include the following ideas.

- Telling people exactly what protective equipment must be worn for each operation and each area, and teaching them how to use the equipment properly. Some companies include in a job description the equipment that must be used on that job.

- Keeping protective equipment stock rooms neat, clean, and orderly. People are reluctant to wear equipment that is issued from a dirty, unkempt stock room. New employees get their first impressions of a company when they are sent to draw equipment, especially safety equipment. If a company does

199

not care about the equipment, how can it care about safety, they reason.

- Keeping stocks of various items. If inventory control is poor and stocks become depleted, it becomes difficult to convince employees that the item is of much value.

- Replacing worn-out equipment promptly. Such equipment is inefficient and may be unsafe. If in good condition, equipment can be reissued after it has been cleaned and sterilized.

Even in companies or departments where personal protective equipment is available, clean, and in good order, many employees do not want to wear it, sometimes in the face of obvious danger. Many people will take off their safety apparel at the first chance they have, when they think they are not being watched.

Although many companies have successful compulsory programs for wearing certain types of equipment, employees must be sold on its value if the program is to be highly successful. This can only be gained by a continued educational program.

"Recognition organizations" for people who have saved life and limb by using safety equipment are listed in Chapter 4, Maintaining Interest, pages 64 and 65.

Paying for protective equipment

Companies differ in their policies for paying for personal protective equipment. Usually companies that require extremely clean operations will furnish laundered coveralls, aprons, smocks, and other garments.

Personal items such as safety shoes and prescription safety glasses, are usually sold on a share-the-cost basis. Some firms maintain safety shoe stores for their employees.

Other items considered necessary to the job are often supplied free to the employee. A welder's helmet and welder's gloves are good examples.

Ordinary work clothes and work gloves are, however, usually purchased by the employee.

It is better economy to assume the cost of necessary equipment than to risk possible injury by not having proper equipment when and where it is needed.

CHAPTER8

Industrial
Housekeeping

There is probably no problem confronting supervision today that is more of a worry and bother than is housekeeping. It is an unending, seemingly thankless chore, that even when well done, usually doesn't stay done. Yet if it is neglected, it seriously interferes with the production and safety goals of a department and of the company.

Why do housekeeping problems occur? Very easily—if top management or the supervisor or both fail to recognize that good housekeeping is one of the vital tools for getting a job done. Like other duties and operations, housekeeping must be planned for and scheduled.

Plan ahead for housekeeping. Leave production floors uncrowded. Provide space for auxiliary tooling, raw materials, or finished product storage; be sure to plan good operating and maintenance procedures.

There always will be some dirt and disorder when you are making a product or providing a service, but if it is allowed to accumulate, sooner or later you will have production problems, higher employee turnover, and increased accident rates. People do not enjoy working in areas that are dirty, disorderly, crowded and booby-trapped with safety hazards. Good housekeeping is good business.

Good housekeeping conditions contribute to production efficiency and safety.

Results of Good Housekeeping

Immediate and long-range results can be obtained by developing and instituting a well planned and well administered housekeeping program.

1. *Reduce operating costs.* Once a department is clean and a housekeeping system has been established, less time and effort are required to keep it clean.
2. *Increase production.* Once obstacles to production are removed, orderly and business-like methods can function without undue interference or delay.
3. *Improve production control.* Material and parts do not get lost or mixed up. Checking operations and recording data are easier.
4. *Conserve materials and parts.* Unused materials, including spoilage and scrap, are easily and quickly removed to the proper place.
5. *Save production time.* There is no need to search for tools, parts or materials.

6. *Better use of floor space.* Clear areas permit free movement of operators and afford repairmen easy access to machines and equipment.

7. *Open aisles* permit faster traffic with fewer collisions.

8. *Lower accident rates.* Sufficient work space and elimination of tripping, slipping, bumping against, being struck by, dropping, and caught between object hazards mean fewer injuries.

9. *Higher employee morale.* Clean surroundings and comparative freedom from injury improve employee attitudes.

10. *Reduce fire hazards.* Fires result from, or are spread by poor housekeeping conditions. Good housekeeping is vital to effective control of fire hazards.

These ten points summarize the main benefits of instituting a good housekeeping program; perhaps you can see even others. What is involved in an effective program is not just a "push broom" effort, but also an orderly arrangement of operations, tooling, equipment, storage equipment, and supplies. A good housekeeping program depends on the planning and scheduling of the housekeeping function into all other operations, and, perhaps above all, a commitment by top management and all supervisors to that end. In the past, some managements have felt so strongly about the problems inherent in poor housekeeping that they have built a good housekeeping policy into their normal operating policies. Perhaps such a policy would help you in your efforts. In some respects, such a policy now exists on the level of law.

Under the Williams-Steiger Occupational Safety and Health Act, each employer is required to maintain a safe and healthful place of employment which is free of all reasonable hazards. More specifically, each of the various subparts of the act usually carry specific housekeeping recommendations for the type of operations they govern—the principal ones being for aisles and passageways, § 1910.22, and materials handling, § 1910.176. Housekeeping during construction operations is specified in § 1926.25. All three sections are reproduced here.

FOR INDUSTRY

§ 1910.22 General requirements.

This section applies to all permanent places of employment, except where domestic, mining, or agricultural work only is performed. Measures for the control of toxic materials are considered to be outside the scope of this section.

(a) *Housekeeping.* (1) All places of employment, passageways, storerooms, and service rooms shall be kept clean and orderly and in a sanitary condition.

(2) The floor of every workroom shall be maintained in a clean and, so far as possible, a dry condition. Where

wet processes are used, drainage shall be maintained, and false floors, platforms, mats, or other dry standing places should be provided where practicable.

(3) To facilitate cleaning, every floor, working place, and passageway shall be kept free from protruding nails, splinters, holes, or loose boards.

(b) *Aisles and passageways.* (1) Where mechanical handling equipment is used, sufficient safe clearances shall be allowed for aisles, at loading docks, through doorways and wherever turns or passage must be made. Aisles and passageways shall be kept clear and in good repairs, with no obstruction across or in aisles that could create a hazard.

(2) Permanent aisles and passageways shall be appropriately marked.

(c) *Covers and guardrails.* Covers and/or guardrails shall be provided to protect personnel from the hazards of open pits, tanks, vats, ditches, etc.

(d) *Floor loading protection.* (1) In every building or other structure, or part thereof, used for mercantile, business, industrial, or storage purposes, the loads approved by the building official shall be marked on plates of approved design which shall be supplied and securely affixed by the owner of the building, or his duly authorized agent, in a conspicuous place in each space to which they relate. Such plates shall not be removed or defaced but, if lost, removed, or defaced, shall be replaced by the owner or his agent.

(2) It shall be unlawful to place, or cause, or permit to be placed, on any floor or roof of a building or other structure a load greater than that for which such floor or roof is approved by the building official.

* * *

§ **1910.176 Handling materials— general.**

(a) *Use of mechanical equipment.* Where mechanical handling equipment is used, sufficient safe clearances shall be allowed for aisles, at loading docks,

through doorways and wherever turns or passage must be made. Aisles and passageways shall be kept clear and in good repair, with no obstruction across or in aisles that could create a hazard. Permanent aisles and passageways shall be appropriately marked.

(b) *Secure storage.* Storage of material shall not create a hazard. Bags, containers, bundles, etc., stored in tiers shall be stacked, blocked, interlocked and limited in height so that they are stable and secure against sliding or collapse.

(c) *Housekeeping.* Storage areas shall be kept free from accumulation of materials that constitute hazards from tripping, fire, explosion, or pest harborage. Vegetation control will be exercised when necessary.

(d) *Drainage.* Proper drainage shall be provided.

(e) *Clearance limits.* Clearance signs to warn of clearance limits shall be provided.

(f) *Rolling railroad cars.* Derail and/or bumper blocks shall be provided on spur railroad tracks where a rolling car could contact other cars being worked, enter a building, work or traffic area.

(g) *Guarding.* Covers and/or guardrails shall be provided to protect personnel from the hazards of open pits, tanks, vats, ditches, etc.

* * *

FOR CONSTRUCTION

§ **1926.25 Housekeeping.**

(a) During the course of construction, alteration, or repairs, form and scrap lumber with protruding nails, and all other debris, shall be kept cleared from work areas, passageways, and stairs, in and around buildings or other structures.

(b) Combustible scrap and debris shall be removed at regular intervals during the course of construction. Safe means shall be provided to facilitate such removal.

(c) Containers shall be provided for the collection and separation of waste, trash, oily and used rags, and other refuse. Containers used for garbage and other oily, flammable, or hazardous wastes, such as caustics, acids, harmful dusts, etc., shall be equipped with covers. Garbage and other waste shall be disposed of at frequent and regular intervals.

✻   ✻   ✻

Supervisory Responsibility

It is up to each supervisor to see that his department area is kept clean and orderly. He must see that housekeeping standards established by top management are maintained, or if there is no program for plant-wide or company-wide housekeeping, then he should set up a program for his own department.

The housekeeping program must be planned and promoted just as carefully and fully as any other activity affecting or requiring participation of all employees. The program must be continuous and procedures set up so operations can be checked frequently and prompt action taken on deviations or a letdown of standards.

The following pages will serve as a guide in setting up such a program.

Clean up

The first step to improve housekeeping is to clean up the department—to remove litter, trash, dirt and accumulated junk. All scrap, salvageable material, broken or unused parts or equipment should be removed.

Getting rid of nonessentials will provide more room to work in. Storage piles will be cleaned up. Personnel will have better facilities and more light. Accidents and materials damage will be reduced. Efficiency will increase, waste will be cut, and esprit de corps will improve.

Layout of equipment, machinery, storage space, aisles, and materials handling facilities, although it affects housekeeping, is a function of production, rather than of housekeeping. As such, it is not within the direct control of the supervisor. How to best utilize available space is, however, a supervisory problem.

A high accident frequency rate is often the direct result of poor housekeeping. Often, supervisors use "after hours" photographs like this to get people to realize just how bad a condition is, and as leverage to get the mess cleaned up.

Directing others

When directing the efforts of other people, and especially when indoctrinating a new employee, the supervisor should make sure that instructions for housekeeping assignments and activities are complete and understood. He should make sure employees understand the advantages they gain through helping maintain orderliness. He should make sure they know that sloppy work practices and cluttered quarters contribute to accidents.

Employee cooperation will most likely be in direct ratio to the efficiency of the system established by the supervisor. Once necessary good housekeeping provisions have been made, the foreman should insist that housekeeping rules be followed.

Here is a list of practices that will help the supervisor secure full cooperation and maintain employee enthusiasm for attaining the maximum degree of orderly housekeeping.

1. Set a good example in your own area. Do not hesitate to pick up unused odds and ends or litter from the floor and put them in the trash can. This impresses people who work for you.
2. Appeal to people's pride. Point out how attractive neat work areas look. Show how each employee can benefit by keeping his work space free from dirt and congestion. Make your program as interesting as possible by giving people a chance to participate, by recognizing efforts, and by conducting an understandable program.
3. Explain specific employee housekeeping responsibilities, why such assignments are necessary, and how they can be carried out.
4. Make sure your instructions are complete and are understood. Leave no chance a person will say later, "I didn't know you meant that."
5. Develop a routine procedure for inspecting the areas in your jurisdiction regularly each week, fortnight, or month.
6. Check closely on general working conditions. It is that first piece of trash on the window sill or under the bench that invites people to add to it. Act immediately when necessary to keep heat, light, ventilation, and sanitation satisfactory.
7. Maintain aisles clear and clean. If storage areas are full, have truckers check with you before unloading.
8. Cooperate with materials handling crews by seeing that temporary storage areas are positively identified and easy to use.
9. Make it easy for people to keep trash off the floor. Make sure trash containers are provided in strategic locations. They must be plainly marked and emptied when full.
10. Make sure proper receptacles are provided for empty milk or beverage cartons or bottles and train employees to use them. Make sure they cooperate with the cleaning crew and with toolroom and stock employees.
11. Permit nothing to be stored on window ledges or hung from walls, even temporarily.
12. Make sure people have seats or benches where needed. Discourage their sitting on kegs or boxes.
13. Check equipment that uses coolants to make sure that oil, coolant, or water does not leak on the floor. See that absorbents are handy for soaking up spilled liquids.
14. Eliminate the practice of keeping excess materials at work places.

This is one of the most prevalent poor work habits.

15. Be sure flammable solvents are kept in approved containers and are used only when needed. Do not permit more than a day's supply to be kept in the department at any time.

16. Encourage employees to report conditions that contribute to disorder.

17. Discourage employee exhibits of "calendar art" that often adorn

The first step to good housekeeping is a good cleanup. Vacuum cleaner is used to collect dust from roof trusses to prevent explosions and/or rapid propagation of fire.

walls and cabinets in work areas. If their removal leaves an obvious mark, you can substitute a safety poster of equivalent size.

18. When investigating accidents or reviewing accident records, determine if faulty housekeeping was a contributing factor.

Stimulating the interest and pride of employees is a normal supervisory function. Special effort is needed, however, when promoting improved housekeeping conditions. Employee behavior and attitudes must be influenced and this job requires help from many sources.

Sources of Help

The supervisor who has special housekeeping problems that he cannot control by himself should have higher management's active support in seeking help from service departments. The supervisor should not charge off with only a nebulous idea of what he wants, but should, rather, first study the machinery, work space, aisles, storage areas, materials requirements, materials flow, and the other factors that contribute to his problems, and then list the solutions to them, so that he can present a clear picture of what is needed.

Safety department

The safety department, or safety professional, has a direct interest in good housekeeping because it affects accident prevention efforts. This department has specialized information on safe handling and storing practices, and also ideas to help plan and promote housekeeping and training efforts.

Maintenance department

Maintenance has direct responsibility for proper functioning of equipment and for the physical condition of building and grounds.

When surveying equipment and facilities, the supervisor should note all repairs needed and then work out a schedule with maintenance. Often the supervisor requisitions small maintenance jobs, and requests approval of his superintendent for major work.

The best maintenance is preventive maintenance, not breakdown maintenance. Preventive maintenance is scheduled as far in advance as possible, as discussed in Chapter 2. The wise supervisor has equipment clean and ready for maintenance by the time the crew arrives.

Janitor service, often a maintenance or plant engineering function, should also be scheduled. Cleaning service should be limited to open areas, such as aisles and central areas. Janitors should not be expected to pick up scrap, spoiled work, processed parts, or tools. Department employees should keep such items off the floor. The supervisor and department clerks are responsible for keeping their immediate work area orderly.

Good lighting—essential to good housekeeping—requires good maintenance.

Planning department

The planning department can help prevent congestion and bottlenecks that make good housekeeping difficult. When the supervisor is responsible for work scheduling, he should confer with supervisors of other departments to set up an efficient work and stock movement schedule in order to prevent congestion.

210

When a staff planning department handles scheduling you should inform them of developments that affect production. Because schedules are based on machine and equipment capacities and on work studies, you must notify the planning department if completion of schedules is hampered by absenteeism, machine breakdown, work spoilage, or any other factor. Planning can then reschedule material arrivals, and can revise or juggle work schedules enough to prevent work from jamming up within the department.

Purchasing department

All materials or equipment coming into your department are usually cleared through and procured by a purchasing agent or department. If requisitioning any new or replacement items, be sure to spell out your requirements for the product. Purchasing agents are not mind readers, but they are trained to buy with economy. Without your specifications to guide them, they may purchase items which will be useless to you and even worse, they could be hazardous to use. Good specifications will enable them to realize economies if a cheaper item really does meet your needs. Good communication is particularly important. When purchases of equipment or materials covered by OSHA regulations are being considered, your familiarity with these requirements will enable you to build these specifications into the purchase order and thus avoid costly modification after installation.

Methods department

Most companies have someone, in addition to the supervisor, who helps improve work methods. Known as methods or process engineering, this group devises ways and means of doing a better job and reducing costs. You should give this group your ideas for improving layout, storage and work areas, and traffic. The methods group, in turn, should suggest ways for making the most of the total space available and for reducing materials handling problems within the department.

Methods engineers can make improvements that might be impossible for a supervisor to make on his own. They can recommend repair or replacement of equipment, make job studies which can improve operations, assist in eliminating obsolete or unused equipment which takes up space, and help the supervisor develop the most practical production flow that requires minimum rehandling.

Plant fire protection

The supervisor should find out from the plant fire protection director, the safety department, or the local fire department what classes of fires might be expected in his area. This depends on the nature of the operations and the character of the materials. Knowing this, make sure that portable fire extinguishers of the correct type and adequate size are available in sufficient quantity. You must then make sure that all employees know where the extinguishers are located and know how to use them.

Inspections

Inspection is to industrial housekeeping as quality control is to production. Inspection does not mean spying, or trying to find something wrong or someone to blame. It is a way of determining whether or not everything is satisfactory, and of uncovering mechanical defects and unsafe plant conditions. Thus it is essential to preventive maintenance as well as housekeeping.

Inspection also reveals many unsafe practices of employees, such as operating equipment without authority or proper training, operating equipment at unsafe speeds, using unsafe materials handling methods, not using proper personal protective equipment, or not wearing proper clothing. When these are found, the supervisor must correct the practices and later follow up to see that they stay corrected.

Techniques

Many supervisors check housekeeping conditions constantly as they go through their department. Others tour their department at the end of their shift. Still others delegate the job to their assistant who, in turn, reports back. When accident records show up housekeeping deficiencies, the inspector should be made aware of the problem and the need to give special attention to certain areas.

Progress chart. Regardless of the inspection method used, a

GOOD HOUSEKEEPING INSPECTION SHEET

AREA __Receiving_____ SUPERVISOR:_____ DATE:_____

DEMERITS:____10_____ RATING:____90.9%_____ BONUS: +0.3%

A—ORDERLINESS, CLEANLINESS
(Cluttered, out of place,
unnecessary, dirty, de-
faced, greasy)

1. Floors, aisles, storage space __B__
2. Trucks, trailers, conveyors _____
3. Desks, files, foremen's
   areas, office _____
4. Corners, out-of-the-way places __+__
5. Machines, furnaces _____
6. Work places, tables, benches _____
7. Tool & supply cupboards or
   containers _____
8. Tool cribs or areas _____
9. Mechanics' benches or areas _____
10. Washrooms, toilets, fountains _____
11. Lockers – personal __+__
12. Yard areas _____

B—SCRAP AND RUBBISH

1. Should have been removed _____
2. No containers _____
3. Wrong type of containers _____
4. Scrap containers not tagged _____
5. Rubbish in scrap containers _____

C—TOOLS AND SUPPLIES

1. Inadequate for purpose __+__
2. Worn out, broken _____
3. No place for _____
4. Wasteful or inefficient use _____

D—MATERIALS

1. Badly piled or blocked _____
2. No ticket or identification _____
3. Should be stocked, scrapped
   or otherwise disposed of _____

E—POWER (Wastage, Leakage)

1. Current _____
2. Steam _____
3. Compressed air _____
4. Heat, fuel, light _____

F—LIGHT AND VENTILATION

1. Condition of fans, blowers,
   hoods, fixtures _____
2. Inadequate light, air,
   ventilation _____
3. Obstructed by dirt, etc. _____

G—MAINTENANCE (Repairs,
   overhauls, replacements)

1. Floors, doors, walls,
   windows _____
2. Wiring, service pipes
   etc _____
3. Machines _____
4. Hoists, tractors, motors _____
5. Other machine accessories _____
6. Cranes, tractors con-
   veyors _____
7. Trucks, trailers __+__
8. Tables stands, benches _____
9. Racks, trays, skids
   platforms __A—__
10. Miscellaneous equipment _____

H—SAFETY

1. Hazard – direct control _____
2. Hazard – indirect control _____
3. Unsafe practice _____
4. Accessibility of stretchers
   fire extinguishers _____
5. Breach of safety rules _____
6. Failure to instruct _____

INSPECTOR_____

Inspection rating sheet can be used as a basis for determining standings in a housekeeping contest, as well as serve as a guide for making inspections. Because of the special conditions and products involved, each company makes its own.

housekeeping progress chart, displayed where all can see it, will keep employees informed and help maintain interest in the housekeeping program.

One supervisor made a floor plan of his department. Work spaces (including his own) had definite boundary lines and names of who was responsible for each area. Employees were encouraged to inspect and help clean their own machines and work areas and to report hazards and bad housekeeping conditions. This, naturally, resulted in compe-

213

tition among employees to maintain the cleanest area and it also relieved the supervisor of much of the work involved in housekeeping and inspection. Employees, because they were involved, also felt part of the whole scheme.

Rating sheet. When inspections are made, a rating sheet should be used. Most companies make their own, which are patterned similarly to the one illustrated.

Rating sheets are essential because they:

- Serve as a guide in making inspections.
- Provide a constant standard for measuring results and progress.
- Remove confusion created by contradictory personal opinions.
- Form a point of reference of evaluating plans and policies.
- Serve as a guide to budgeting and scheduling.
- Rule out individual whims and prejudices.
- Provide a fair basis for awarding recognition.

When making an inspection, the supervisor should pay particular attention to the following hazard areas. In addition, all of these areas are covered by OSHA regulations which may detail specific instructions for compliance. To simplify conformance to OSHA regulations, the National Safety Council publishes two "OSHA Standards Checklists," one for occupational (part 1910) and the other for construction (part 1926). The inspection sheets are printed only on one side in order to be used on a clipboard.

Plant structure

Walls and columns should not be used as places to hang clothing, job tickets, calendars and pinup girls, rags, or similar objects. Once one item is up, it invites others and soon the wall is covered.

Window sills are not storage spaces. Even one milk carton left on a ledge is just the start toward making the ledge a catchall.

Shelves should be kept in repair. They are generally not strong enough to be used for storage. Even if they are, a person could be struck by a falling object.

Stairs and exits should be well lighted and designated as exits. They should be free from obstructions, have treads and handrails in good repair, and should be kept clean. Even a small bolt or pencil could cause a bad injury, if dropped and not picked right up.

Clearly defined aisles, wide enough to handle safely the type and amount of traffic, must be kept clean. Sweepers used in areas with low-overhead obstructions should have metal-frame overhead guard.

Working platforms 4 ft or more high must have strong guardrails, and also toeboards if there is a danger of material falling off.

Floors

Floors should be level and kept as slip-proof as possible. They should be free of imbedded chips and accumulated drippings. Material that might drip or spill can be collected in drip pans, or gutters, or be deflected by splash guards. If these liquids do get on the floor, nonflammable absorbent materials should be readily available. Sawdust, because of its combustibility, should never be used as an absorbent.

215

Floors that are not swept or vacuumed as often as the dustiness of the operation demands should be reported. Cracks, splinters, ruts, and breaks in the floor should be repaired as soon as they are discovered. Maintenance must be proper for the type floor or floor covering in the various areas.

Safe load limits for floors must be determined, marked, and maintained.

Aisles

The type and amount of traffic that uses an aisle determines how wide it should be. The minimum width is that required by four men carrying a stretcher.

Aisles should be clearly defined by painted lines, plastic tape, guard rails, or some other satisfactory means. These also serve as a constant reminder that aisles are to be kept clear. The portion of the aisle used for pedestrian traffic should be clearly designated.

Blind corners and intersections should have (a) nonshattering mirrors placed so that traffic in an aisle is visible to anyone approaching from the side, and (b) warning signs to remind all traffic that it is approaching a corner or intersection. All traffic should be kept moving.

Bump rails should be installed along shop offices, storage areas, and machinery as protection from trucks. Aisles should not be used as storage space.

Storage facilities

Storage space should be marked off and kept separate. Its location and size should be realistic and contribute to the efficiency of the department. Incoming and outgoing materials should be kept separate. Departmental space should not be used for long-term storage of small parts, such as nuts, bolts, washers, and springs.

Piling, stowing, or stocking should be safe, orderly, and neat. Piles should be kept below sprinkler heads at a distance prescribed by fire regulations. See page 387.

If materials handling equipment is kept in the department, a storage space should be allotted for it also. Containers should never be overloaded and should be kept in good repair.

Floor loads, approved by the building official, shall be marked on plates which shall be supplied and affixed by the owner of the building, or his agent, in a conspicuous place in each space to which they shall relate. Where the same type of material is stored regularly, walls and columns can be marked to indicate the height to which material can be piled without exceeding the allowable floor load.

Cabinets used for storage should be kept closed and the permissible contents labeled on the outside. Other material should be kept out. Flammable material should be stored in a separate area or building as prescribed by fire regulations.

Racks are considered the best solution for storing ladders. Never leave a ladder on the floor or against a wall or partition.

Employee facilities

Personal belongings belong in lockers, not in work areas. Lockers should be cleaned out and inspected once a week to prevent unhealthful or unsanitary accumulations.

Toilet rooms should be well ventilated, well lit, and provided with separate receptacles for disposal of towels and smoking materials. Floors, toilets, and lavatories should be cleaned at least once each shift, and more often if necessary. Lavatories should have hot and cold

217

water, soap, and towels. Sterilizers or deodorizing disinfectants should be used, but it is better to eliminate the source of odors rather than cover them up with strong-smelling disinfectants.

Cleaning and sanitation material and equipment should be stored

Make it easy for employees to dispose of personal trash properly.

carefully to prevent waste. Such equipment should not be stored in toilet or locker rooms, in halls, in corners, or on stairways.

Eating space, with adequate bottle and trash receptacles should be provided and cleaned at least once a shift. Employees should not be allowed to litter the floor. Drinking fountains and any beverage- or food-dispensing facilities should be cleaned each shift.

Cuspidors should be provided in work areas for those who chew tobacco or use snuff.

218

Smoking must be strictly prohibited when it could be a fire hazard. "No smoking" signs must be posted. Both eating and smoking should also be prohibited in areas containing toxic materials that could enter the mouth by these means.

Machines and equipment

Like aisles and storage areas, machine areas should be marked off, and machines placed so that operators are not exposed to aisle traffic. Allow space for maintenance and repair, for storage of incoming and outgoing processed stock, and for tools, jigs, and fixtures needed. Tools, jigs, and fixtures are not to be left around haphazardly, but kept in racks and containers.

Work benches are not catchalls for personal belongings, fixtures, tools, spare parts, and odds and ends. They should be kept clean, and in good repair. Good lighting, tool racks, cabinets for extension cords and air hoses, and storage space for dies and gages and other equipment should be provided. Trash containers should be spotted conveniently throughout the work area.

Scrap, clips, cuttings, and dust should be dumped directly from machines or benches into containers and not onto the floor. Containers should be emptied regularly. Spoiled work should be taken for reclamation at least twice a shift. Finished or in-process items should be fed and taken away efficiently and safely and not left on the machine or on the floor. Both scrap and work containers should be rated for specified loads and removed promptly when full.

Proper lubrication of machinery will make housekeeping easier. Follow manufacturer's instructions for lubrication. Train and supervise oilers to avoid using too much oil or grease, and to clean up any spilled oil or grease before they leave a machine or area. Not only does this make the area look neater, but it helps prevent falls and fires.

Boxes and kegs are not designed to substitute as seats. If an operator must sit down, a proper chair should be provided.

Machines should be kept clean and in proper adjustment and repair. These should be made promptly, as discussed under "Maintenance Department," earlier in this chapter. Machines should be painted colors that make it as easy as possible to see the work, spotlight the point of operation, and service. (Details are in a subsequent section "Color.")

219

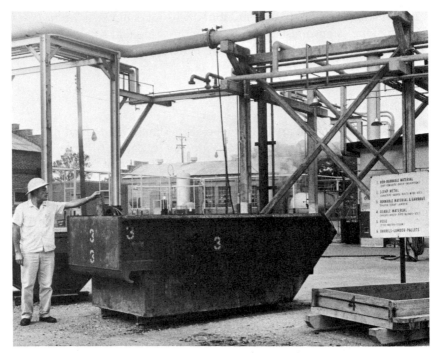

"A place for everything, and everything in its place" is truly the secret of good housekeeping. Separate waste containers for different classes of scrap help keep this yard clean as well as aid salvage. Valuable items are not lost. Unusable items are disposed of properly, not laid aside as stumbling blocks.

High-hazard areas should be clearly marked by signs that must be kept clean and legible at all times. Obsolete signs should be removed. Machine safeguarding is discussed in Chapter 10.

Yards and grounds

Yards and grounds should be carefully maintained. Follow the same principles recommended for inside storage. Keep bulk materials in neat, well trimmed piles; keep grass and weeds cut down.

Traffic ways and railway spurs should be kept free of trash and obstructions, and be kept well drained to reduce ice and mud and the chances of skids, slips, or other mishaps. (Keep abrasive or ice-melting material handy, just in case.)

Traffic lines effectively control vehicle and pedestrian movement.

Color

Industry uses color in four ways:

- For plant interiors and general equipment,
- For marking hazards and identifying particular equipment,
- For identifying content of piping systems, and
- For accident prevention signs.

The safety uses are the main concern of this book.

Marking hazards and protective equipment. The *Safety Color Code for Marking Physical Hazards,* American National Standard Z53.1, provides for uniform marking of physical hazards and to identify and indicate the location of fire and other protective equipment. Use of color is not a substitute for safeguards, but rather is a supplement to them.

The Standard specifies the following color applications.

- *Red—Fire, danger, stop*

Identifies fire protection equipment and apparatus including exit signs.

Wall marking and supports on which extinguishers are mounted.

Flammable liquid containers (except shipping containers), on which the name of the contents should be stenciled.

Danger signs and lights at barricades, temporary obstructions, and temporary construction sites.

Emergency stop bars on machines like rubber mills, wire blocks, and flat work ironers; stop buttons for electric switches.

- *Yellow—Caution for marking physical hazards*

Identifies hazards that may result in slipping, falling, tripping, caught in-between, and bumping into objects.

For attracting attention—solid yellow, yellow with black stripes, or yellow and black checks. (Use the combination that creates the most attention.)

Handrails, guard rails, or top and bottom treads of stairways, low beams, pipes, and crane blocks. Exposed edges of platforms, pits, and walls.

Mobile equipment (construction or materials handling) and industrial locomotives—black and yellow stripes.

- *Green—Safety and first aid equipment location*
  First aid and safety equipment (like deluge showers).
  Stretchers, gas masks, and safety bulletin boards.
  Other locations connoting "safety" rather than "hazard."

- *Black and white—Traffic and housekeeping*
  Housekeeping and traffic markings, in either stripes, checks, or solid color.

- *Orange—Dangerous parts of machines and electrical equipment*
  Dangerous parts of machines that are exposed when enclosure doors are open or guards removed.
  Energized equipment which may cut, crush, shock, or otherwise injure.
  Inside of movable guards.
  Safety starting buttons.
  Exposed edges of cutters, gears, pulleys, sheaves, and the like.

- *Blue—Operation warnings*
  Barriers, flags, tags, and other cautions against starting, using or moving machinery or equipment undergoing repair or being serviced. These should be placed conspicuously at the starting points or power sources or valves of machinery, such as elevators, tanks, ovens, mixers, and boilers, and on scaffolds and ladders.
  The 1971 standard recommends that yellow replace blue.

- *Purple—Radiation hazards*
  Containers of radioactive materials.
  Rooms, areas, or equipment that is contaminated with radioactive materials.
  Radioactive materials disposal and storage areas.
  Signal lights that indicate radiation-producing machines in operation.
  Yellow is used in combination with purple for tags, labels, signs, and floor markers.

The color code for piping systems (See Table VIII) is a simplified plan to classify and identify pipeline contents. The basic colors are much the same as those in the *Safety Color Code, Z53.1.* This piping code does not apply to pipes buried underground, to electrical conduit, or to pipes carrying solids suspended in either gas or air.
The value of coding is principally for maintenance. In processing

TABLE VIII

IDENTIFICATION OF PIPELINE CONTENTS BY COLOR

Material Classification	Color
Fire protection materials.............	red
Dangerous materials..................	yellow
	orange
	green
	white
Safe materials......................	black
	gray
	aluminum
Protective materials.................	bright blue

Source: American National Standard A13.1, *Scheme for the Identification of Piping Systems.*

and warehouse areas, the color may be applied to the full length of the pipe, or in bands 8 to 10 in. wide near valves and pumps and repeated at intervals along the line. For fire sprinkler lines in offices, painting where the line enters and leaves a room (or penetrates a partition) is sufficient for most local codes and keeps the room from looking like it is decorated for a holiday.

The name of the specific material in a line can be stenciled in black or a sticker bearing the name can be placed on the pipe at easily visible locations, such as valves, pumps, and passes through walls or partitions. Contents of pipes less than ¾-in. diameter can be identified by enameled metal tags.

The color code for accident prevention signs follows closely the colors for the two uses just discussed. The advantage is that persons unable to read can sense the cautioning signals of the signs through color association. The code designates not only the color but also the size and lettering for each of several kinds of signs. See § 1910.145 for OSHA specifications for accident prevention signs and tags.

Maintenance is important. Dust and dirt accumulations on walls and ceilings, windows and skylights, and lighting fixtures reduce illumination. Faded, washed-out paint also reduces the protective value of contrasting colors on machinery, identifying colors of equipment, and distinctive colors on signs. When these conditions develop, the supervisor should report them to the proper authority for cleaning and repainting as necessary.

CHAPTER 9

Materials Handling and Storage

Materials handling is a job that almost every worker in industry performs—either as his sole duty or as part of his regular work, either by hand or with mechanical help.

Manual handling of materials increases the possibility of injuries and adds to the cost of a product. To reduce the number of materials handling injuries and to increase efficiency, materials handling should be minimized by combining or eliminating operations. What materials handling must still be done should be done mechanically, insofar as possible.

In many industries, materials could not be processed at low cost if it were not for mechanical handling. Although mechanized handling creates a new set of hazards, the net result (entirely aside from increased efficiency) is fewer injuries.

Materials handling problems

Manual handling of materials accounts for about 23 percent of all occupational injuries—these injuries are from every part of an operation, not just the stock room or warehouse. As an average, industry moves about 50 tons of material for each ton of product produced. Some industries move 180 tons for each ton of product.

Strains and sprains, fractures, and bruises are the common injuries. They are caused, primarily, by unsafe work practices—improper lifting, carrying too heavy a load, incorrect gripping, failing to observe proper foot or hand clearances, and failing to wear proper equipment.

To gain insight on your materials handling injury problem, you should ask these questions about your present operating practices.

1. Can the job be engineered so that manual handling will not be necessary?
2. How do the materials being handled (such as chemicals, dusts, rough and sharp objects) hurt the people doing the handling?
3. Can employees be given handling aids, such as properly sized boxes, adequate trucks, or hooks, that will make their jobs safer?
4. Will protective clothing, or other personal equipment, help prevent injuries?

These are by no means the only questions that might be asked, but they serve as a start toward overall appraisal and detailed inquiry.

Manual Handling Methods

The largest number of injuries occur to fingers and hands. People need instruction if they are to avoid the "natural" ways of picking objects up and putting them down. You can give them these pointers.

1. Inspect materials for slivers, jagged edges, burrs, rough or slippery surfaces.
2. Get a firm grip on the object.
3. Keep fingers away from pinch points, especially when setting down materials.
4. When handling lumber, pipe, or other long objects, keep hands away from the ends to prevent them from being pinched.
5. Wipe off greasy, wet, slippery, or dirty objects before trying to handle them.
6. Keep hands free of oil and grease.

In most cases, gloves, hand leathers, or other hand protectors have to be used to prevent hand injuries. Their use must be controlled if they are to be worn around moving machinery. (See details in Chapter 7.)

Correct gloves or hand protection will reduce the number of injuries to fingers and hands—by far the most frequently injured parts of the body.

In other cases, handles or holders can be attached to objects themselves, such as handles for moving auto batteries, tongs for feeding material to metal-forming machinery, or wicker baskets for carrying control laboratory samples.

Feet and legs sustain a share of materials handling injuries, the greatest share being to the toes. One of the best ways to avoid injuries is to have people wear foot protection—safety shoes, instep protectors, and ankle guards.

Eyes, head, and trunk of body can also be injured. When opening a wire-bound or metal-bound bale or box, a person should wear eye protection as well as stout gloves, and take special care to prevent the ends of the bindings from flying loose and striking his face or body. The same precaution applies to coils of wire, strapping, or cable.

If material is dusty or is toxic, the person handling it should wear a respirator or other suitable personal protective equipment. (See Chapter 7.)

People also need training in handling heavy objects. See the following directions for lifting and carrying.

Lifting and carrying

Before employees are assigned to jobs requiring heavy and/or frequent lifting, make sure they are physically suited for the job. A person's lifting ability is not necessarily indicated by his height or weight. In some cases a small man can lift heavy objects safely, whereas a big, husky 250-pounder may not be able to do so without injuring himself. Some state laws limit the load a worker can lift. Because these limits vary widely, the supervisor should know his state law and abide by it. Some states have laws that apply to how much a woman may lift in continuous or repeated operations. These must be followed also.

If a load is thought to be more than one person should handle, assign more than one employee to the operation or supply materials handling equipment.

Be sure to instruct properly. If he knows how, a man can lift loads of considerable weight. If he does not know how, he may strain himself when lifting a comparatively light load.

Lifting and setting down are the first and last movements performed in handling materials. When done by hand, it is during these movements that most strains occur. It is important to train employees in proper lifting techniques if lifting injuries are to be reduced.

1. Consider the size, weight, and shape of the objects to be carried. Do not lift more than can be handled comfortably. If necessary, get help.
2. Never carry a load that you cannot see over or around. Make sure the path of travel is clear.
3. Follow the 6-step lifting procedure described on pages 228–229.
4. Setting down an object requires just the reverse procedure.

When bulky objects are to be handled or when objects are to be carried on the shoulder, train your people in these techniques for specific situations.

1. To place an object on a bench or table—First set the object on edge and push it far enough onto the support to be sure it will not fall. Release it gradually as you set it down. Move it in place by pushing with the hands and body from in front of the object. This method prevents fingers from getting pinched.
2. To raise an object above shoulder height—Lift the object first to waist height. Rest the edge of the object on a ledge, stand, or hip.

Proper Way To Lift

Lifting is so much a part of everyday jobs that most of us don't think about it. But it is often done wrong, with bad results: pulled muscles, disk lesions, or painful hernia.

Here are six steps to safe lifting.

1. Keep feet parted—one alongside, one behind the object.
2. Keep back straight, nearly vertical.
3. Tuck your chin in.
4. Grip the object with the whole hand.
5. Tuck elbows and arms in.
6. Keep body weight directly over feet.

FEET should be parted, with one foot alongside the object being lifted and one behind. Feet comfortably spread give greater stability; the rear foot is in position for the upward thrust of the lift.

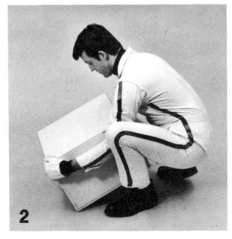

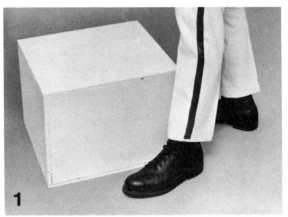

BACK. Use the sit-down position and keep the back straight —but remember that "straight" does not mean "vertical." A straight back keeps the spine, back muscles, and organs of the body in correct alignment. It minimizes the compression of the guts that can cause hernia.

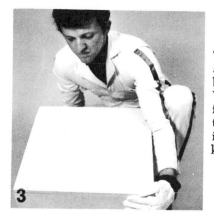

3

Arms and elbows.
The load should be drawn close,
and the arms and elbows should
be tucked into the side of the body.
When the arms are held away
from the body, they lose much of
their strength and power. Keeping the arms tucked in also helps
keep body weight centered.

Palm. The palmer grip
is one of the most important elements
of correct lifting. The fingers and
the hand are extended around the object you're going to lift. Use the full
palm; fingers alone have very little
power. Glove has been removed to
show finger positions better.

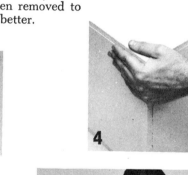

4

5

Chin. Tuck in
the chin so your neck and
head continue the straight
back line and keep your
spine straight and firm.

Body weight. Position
body so its weight is centered over
the feet. This provides a more
powerful line of thrust and ensures
better balance. Start the lift with
a thrust of the rear foot.

6

Shift hand position so object can be boosted after knees are bent. Straighten out knees as object is lifted or shifted to the shoulders.

3. To change direction—Lift the object to carrying position, and turn the entire body, including the feet. Do not twist your body. In repetitive work, the person and the material should both be positioned so that the person will not have to twist his body when moving the material.

4. If the object is too heavy to be handled by one person—Get help. When two or more men are handling the same object, one man should "call the signals." All the men on the lift should know who this man is and should warn him if any one of the crew is about to relax his grip.

Handling specific shapes

Boxes or cartons are best handled by grasping them at alternate top and bottom corners and drawing a corner between the legs.

Bags or sacks are grasped in the same manner as boxes. If a sack is to be raised to shoulder height, it should be raised to waist height first and rested against the belly or hip before it is swung to one shoulder so that it rests on its side.

Barrels and drums. Men need special training to handle barrels and drums safely.

If two men are assigned to up-end a full drum, they should use the following procedure.

1. Stand on opposite sides of the drum and face each other.
2. Grasp both chimes (rolled edges at both ends of the barrel) near their high points. Lift one end; press down on the other.
3. As the drum is upended and brought to balance on the bottom chime, release the grip on the bottom chime and straighten up with the drum.

When two men are to overturn a full drum, they use this procedure.

1. Make sure they have enough room. Cramped quarters can result in badly injured hands.
2. Both stand near each other, facing the drum. They grip the closest point of the top chime with both hands. Resting their palms against

230

Two-handled lifter minimizes strain and provides safe control when lifting or lowering a drum.

the side of the drum, they push until the drum balances on the lower chime.

3. They step forward a short distance, and each man releases one hand from the top chime in order to grip the bottom chime. They ease the drum down to a horizontal position until it rests solidly on its side.

If one man is to overturn a drum, he should:

1. Make sure he has enough room.
2. Stand in front of the drum, reach over it, and grasp the far side of the top chime with both hands. (A short man can grasp the near side of the chime, if this is easier.) If the drum is tight against a wall or against other drums, pull on the chime with one hand and

Drop-bed truck design, by eliminating high lift of soft drink cases, reduces materials handling requirements by about 30 percent.

push against the wall (or other drum) with the other hand for additional control.

3. Pull the top of the drum toward him until it is balanced on the edge of the lower chime.

4. Transfer both hands to the near side of the top chime. Keep hands far enough apart to avoid their being pinched when the drum touches the floor.

5. Lower the drum. Keep the back straight, inclined as necessary. Bend legs so that the leg muscles take the strain.

If one man is to upend a drum, he reverses this procedure. Actually, if one man must handle a drum, he should have a lifter bar which hooks over the chime and gives him powerful leverage and excellent control. Barrel and drum lifters are commercially available.

To roll a barrel or drum, a man should push against the sides with his hands. To change direction of the roll, he should grip the chime, not kick the drum with his feet.

To lower a drum or barrel down a skid, turn it and slide it endwise. Do not roll it. To raise a drum or barrel up a skid, two men stand on opposite sides of the skid (outside the rails, not inside, and not below the object being raised). They roll the object up the incline.

Handling drums and barrels can be hazardous, even when using utmost care. Special handling equipment and tools (approved by the safetyman or safety department) should be made available to make the job safer and easier.

Long objects, like ladders, lumber, or pipe, should be carried over the shoulder. The front end should be held as high as possible to prevent its striking other employees, especially when turning corners. When two or more carry an object, they should place it on the same shoulder, respectively, and keep in step.

Equipment for Handling

Handling materials with the hands alone is usually slow and hazardous. To make handling safe and efficient, much equipment has been developed—from small hand tools to power trucks and conveyors.

Hand tools

Many hand tools are available for specific jobs or for specific materials. They should only be used for their specific purpose.

The crowbar is probably the most common hand tool. Select the proper kind and size for the job. Because the bar can slip, a man should never work astride it. He should so position himself that he will not be pinched or crushed if the bar should slip or the object move suddenly. The bar should have a good "bite" and not be dull or broken. When not in use, it should be hung on a rack or otherwise placed so it cannot fall on or trip someone.

Rollers are used for moving heavy or bulky objects. Men should be careful not to crush their fingers or toes between the rollers and the floor. They should use a sledge or bar, not their hands and feet, to change direction of the object. They should avoid unnecessary turns by making sure the rollers are properly placed before moving is started. Never use compressed gas cylinders as rollers.

Handles of tongs and pliers should be offset so the user will not pinch his hands and fingers.

Hooks. Hand hooks used for handling material should be kept sharp so that they will not slip when applied to a box or other object.

The handle should be strong and securely attached and shaped to fit the hand. The handle and the point of long hooks should be bent on the same plane so that the bar will lie flat when not in use and not constitute a tripping hazard. The hook point should be shielded when not in use.

Shovels. Shovel edges should be kept trimmed, and handles should be checked for splinters. Workers should wear safety toe shoes with sturdy soles. They should have their feet well separated to get good balance and spring in the knees. The leg muscles should take much of the load.

To reduce the chance of injury, the ball of the foot—not the arch —should be used to press the shovel into clay or other stiff material. If the instep is used and the foot slips off the shovel, the sharp corner of the shovel may cut through the workman's shoe and into the foot.

Dipping the shovel into a pail of water occasionally will help to keep it free from sticky material, making it easier to use and less likely to cause strain. Greasing or waxing the shovel blade will also

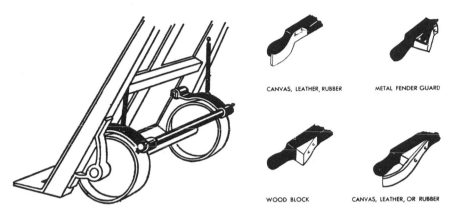

CANVAS, LEATHER, RUBBER   METAL FENDER GUARD

WOOD BLOCK   CANVAS, LEATHER, OR RUBBER

Hand truck safety features. Photo (on left page) shows recessed handle that makes certain that hands are not exposed to pinch points between truck and walls or other trucks. (Wooden truck is for service in an explosives plant.) Drawings show: (left above) a two-wheeled truck equipped with foot brakes to keep wheels from slipping when truck is being loaded; (right) various types of knuckle guards.

prevent some kinds of material from sticking. When not in use, shovels should be stood against a wall or kept in racks or boxes.

Hand trucks

Two-wheeled hand trucks look as though they should be easy to handle, but there are safe procedures that must be followed.

1. Keep the center of gravity of the load as low as possible. Place heavy objects below lighter objects.
2. Place the load well forward so the weight will be carried by the axle, not by the handles.
3. Place the load so it will not slip, shift, or fall. Load only to a height that will allow a clear view ahead.
4. Let the truck carry the load. The operator should only balance and push.
5. Never walk backwards with a hand truck.
6. When going down an incline, keep truck ahead. When going up, keep truck behind.
7. Move trucks at a safe speed. Do not run. Keep truck constantly under control.

A truck designed for a specific purpose should only be used for that purpose—a curved bed truck should only be used for handling drums or other circular materials, and a horizontal platform truck for handling acetylene or compressed gas cylinders.

Foot brakes may be installed on wheels of two-wheeled trucks so that operators need not place their feet on the wheel or axle to hold the truck.

Handles should have knuckle guards.

Four-wheeled truck operation follows rules similar to those for two-wheeled trucks. Extra emphasis should be placed on proper loading, however. Four-wheeled trucks should be evenly loaded to prevent their tipping. Four-wheeled trucks should be pushed rather than pulled, except for a truck that has a fifth wheel and a handle for pulling.

Trucks should not be loaded so high that operators cannot see where they are going. If there are high racks on the truck, two men should move the vehicle—one to guide the front end, the other to guide the back end. Handles should be placed at protected places on the racks or truck body so that passing traffic, walls, or other objects will not crush or scrape the operator's hands.

Truck contents should be arranged so they will not fall or be damaged in case the truck or the load is bumped.

General precautions. Truckers should be warned of three main hazards: (a) running wheels off bridge plates or platforms, (b) colliding with other trucks or obstructions, and (c) jamming their hands between the truck and other objects.

When not in use, trucks should be stored in a designated area, not parked in aisles or other places where they would be a tripping hazard or traffic obstruction. Trucks with drawbar handles should be parked with handles up and out of the way. Two-wheeled trucks should be stored on the chisel with handles leaning against a wall or the next truck.

Powered hand trucks

The use of the powered hand truck, controlled by a walking operator, is increasing. The principal hazards of this truck are (a) the operator can be caught between the truck and another object, and (b) collisions with other objects.

Lead a powered hand truck. Face direction of travel. Keep one hand on the handle. Always give pedestrians the right of way.

The truck should be equipped with a dead-man control, wheel guards, and an ignition key that can be taken out when the operator leaves the truck.

Most operators *must* be trained and not use trucks unless authorized. Training should include operating instructions given in the truck manufacturer's manual. General instructions include:

1. Do not operate the truck with wet or greasy hands.
2. Lead the truck from right or left of the handle. Face direction of travel. Keep one hand on the handle.
3. When entering an elevator, back the truck in to keep from getting caught between the handles and the elevator walls. Operate the truck in reverse whenever it must be run close to a wall or other obstruction.
4. Always give pedestrians the right of way.
5. Stop at blind corners, doorways, and aisle intersections to prevent collisions.
6. Never operate the truck faster than normal walking pace.

Trucks designed for riding should be equipped with a strong platform extending beyond the operator's position.

7. Only handle flammable or corrosive liquids when they are in approved containers.
8. Never ride the truck, unless it is specifically designed for the driver to ride.
9. Never permit others to ride on the truck.
10. Do not indulge in horseplay.

These general rules for operators of powered hand trucks should also be followed by operators of power trucks.

Powered industrial trucks

Only physically qualified operators, who have received training in safe operation and who are duly authorized, are permitted to drive any industrial power truck.

According to OSHA requirements, training programs should include safe operating practices, as well as actual supervised experience driving over a training course. Emphasis should be on safety awareness. Trained and authorized drivers should have badges or other visual identification of authorization to drive, and they should display these at all times.

Power trucks have either a battery-powered motor or an internal combustion engine. Trucks should be maintained according to their manufacturer's recommendations. All trucks acquired on or after February 15, 1972, must meet the design and construction requirements established in the *American National Standard for Powered Industrial Trucks*, Part II, B56.1–1969. Modifications and additions which affect capacity and safe operation shall not be performed by the customer or user without the manufacturer's prior written approval. All name-plates and markings must be accurate, in place, and legible. Eleven different designations of trucks or tractors are authorized for use in various locations. For details, see OSHA regulation § 1910.178, which is promulgated from ANSI B56.1–1969 and NFPA No. 505–1969.

Battery-charging installations must be located in areas designated for that purpose. They must have facilities for flushing and neutralizing spilled electrolyte, for fire protection, for protecting charging apparatus from damage by trucks, and for adequate ventilation for dispersal of gases or vapors from gassing batteries. Racks used for supporting batteries must be made of materials nonconductive to spark generation or be coated or covered to achieve this objective.

A conveyor, overhead hoist, or equivalent equipment must be used for handling batteries. Reinstalled batteries must be properly positioned and secured in the truck.

A carboy tilter or siphon must be used for handling electrolyte. Acid must always be poured into water; water must not be poured into acid (it overheats and splatters).

During charging operations, vent caps must be kept in place to avoid electrolyte spray. Make sure that vent caps are functioning. Battery or compartment cover or covers must be open to dissipate heat.

Precautions must be taken to prevent open flames, sparks, or electric arcs in battery-charging areas and tools and other metallic objects must be kept away from the tops of uncovered batteries.

Employees charging and changing batteries must be authorized

239

Batteries and charging apparatus are protected from damage by trucks.

to do the work, trained in the proper handling and required to wear protective clothing, including face shields, long sleeves, rubber boots, aprons, and gloves.

Smoking is prohibited in the charging area.

Refueling. All internal combustion engines must be turned off before refueling. Refueling should be in the open or in specifically designated areas, where ventilation is adequate to carry away fuel vapors. Liquid fuels, such as gasoline and diesel fuel, must be handled and stored in accordance with the National Fire Prevention Association *Flammable and Combustible Liquids Code,* No. 30–1969 (now OSHA regulation § 1910.106); liquefied petroleum gas fuel (or LP-gas), in accordance with the National Fire Prevention Association's *Handling of Liquefied Petroleum Gases,* No. 58–1969 (now OSHA regulation § 1910.110).

Smoking must not be permitted in the service areas and signs must be posted to that effect.

General rules for driving powered industrial trucks in most operations are as follows. (For each operation, a specific set of rules should be drawn up.)

1. All traffic regulations must be observed, including authorized plant speed limits.
2. Safe distances must be maintained. Approximately three truck lengths from the truck ahead and truck must be kept under control at all times, so that an emergency stop, if necessary, can be made in the clear distance ahead.
3. The right of way shall be yielded to ambulances, fire trucks, and other vehicles in emergency situations.
4. Other trucks traveling in the same direction must not be passed at intersections, blind spots, or other dangerous locations.
5. Drivers are required to slow down and sound horns at cross aisles and other locations where vision is obstructed. If forward view is obstructed, because of the load being carried, drivers are required to travel with the load trailing.
6. Railroad tracks must be crossed diagonally whenever possible, and parking within 8 ft of the center of the nearest railroad track bed is prohibited.
7. Drivers are required to look in the direction of travel, and to keep a clear view of the path of travel at all times. Never back up without looking.
8. Grades are to be ascended or descended slowly and loaded trucks must be driven with the load upgrade on grades in excess of 10 percent. Unloaded trucks must be operated on all grades with load engaging means downgrade. On all grades, load and load-engaging means shall be tilted back if applicable and raised only as far as necessary to clear the road surface.
9. Under all travel conditions, truck must be operated at a speed that will permit it to be brought to a stop in a safe manner. Drivers are required to slow down for wet and slippery floors and stunt driving and horseplay are not to be permitted.
10. Dockboards or bridgeplates are to be driven over carefully and slowly and only after they have been properly secured. Their rated weight capacity must never be exceeded.

11. Approach elevators slowly and stop at least 5 ft from gate. Obey elevator operator's signals, enter squarely and only after the elevator is properly leveled. Once on the elevator, neutralize the controls, set the brakes, shut off power, and then step off the truck.

12. While negotiating turns, speed must be reduced to a safe level by means of turning the hand steering wheel in a smooth, sweeping motion. When maneuvering at a very low speed, the hand steering wheel must be turned at a moderate even rate.

13. Never run over loose objects on the roadway surface.

14. Only stable or safely arranged loads shall be handled and caution must be exercised when handling off-center loads which cannot be centered.

15. Only loads within the rated load capacity of the truck shall be handled, and long or high (including multiple-tiered) loads which may affect capacity must be adjusted.

16. Load-engaging means must be placed under the load as far as possible and the mast shall be carefully tilted backward to stabilize the load.

17. Extreme care must be exercised when tilting the load forward or backward, particularly when high tiering. Tilting forward with load-engaging means elevated is not permitted except to pick up a load. Elevated loads must not be tilted forward except when the load is in a deposit position over a rack or stack. When stacking or tiering, only enough backward tilt to stabilize the load shall be used.

18. When operating in close quarters, keep hands where they cannot be pinched between steering controls and projecting stationary objects. Keep legs and feet inside the guard or the operating stations of the truck.

19. Do not use the reverse control on electric trucks for braking.

20. Park trucks only in designated areas—never in an aisle or doorway, or obstructing equipment or material. Fully lower the load engaging means, neutralize the controls, shut off the power, and set the brakes. Remove the key (or connector plug) when leaving a truck unattended. If the truck is parked on an incline, block the wheels.

Depending on its use, a power truck should have a dead-man control, and should be equipped with guards, such as hand enclosure guard or a canopy guard. A lift truck should also have upper and

Overhead guard for forklifts. Bars run parallel to driver's line of vision so no cross member gets in the way of his view as he maneuvers high pallets.

lower limit switches to prevent overtravel. No power truck should be used for any purpose other than that for which it has been designed.

A forklift should not be used as an elevator for employees (for example, in servicing light fixtures or when stacking materials), unless a safety pallet, with standard railing and toeboards, is fastened securely to the forks.

When internal combustion engine trucks are operated in enclosures, the concentration of carbon monoxide should not exceed the limits specified by local or state laws, and in no case should the time-weighted average concentration ever exceed 50 ppm (parts per million) for an 8-hour exposure. The atmosphere must contain a minimum of 19 percent oxygen, by volume. (Air usually contains 20.8 percent oxygen, by volume.)

Many companies equip their fuel-powered trucks with catalytic exhaust purifiers to burn the carbon monoxide before it ever reaches the atmosphere.

Some power trucks are designated for use in locations where flammable gases or vapors are or may be present in the air in quantities sufficient to produce explosive or ignitable mixtures, or which are hazardous because of the presence of combustible dust or easily ignitable fibers or flyings. Only trucks approved and marked for such areas are to be permitted in them. (See OSHA regulations § 1910.178 or NFPA Standard No. 505, *Powered Industrial Trucks,* for classification details.)

Loads. Operators should be told to refuse unstable loads. If material of irregular shape must be carried, it should be placed so that it will not fall off, and be blocked, tied, or otherwise secured. A load should, if possible, be cross-tied after being neatly piled. An extra counterbalance should never be placed on a forklift to facilitate handling overloads, because it puts added strain on the truck and endangers the operator and nearby employees. It is hard to operate an overloaded truck safely.

Operators of powered industrial trucks must not drive up to anyone standing in front of a bench or other fixed object, or allow anyone to stand or pass under the elevated portion of any truck, whether loaded or empty. Unauthorized personnel are not permitted to ride on trucks; even if authorized, then they are permitted only when they are in the place provided. The placing of arms or legs between uprights of the mast or outside the running lanes of the truck, by operators or authorized riders is prohibited.

Raise or lower loads at the point of loading or unloading, not during travel. Operators should be trained to look before raising a load so that they will not strike structural members of the building, electric wiring or cables, or piping (especially that carrying gases or flammable fluids).

Operators must make sure there is sufficient headroom under overhead installations, such as lights, pipes, and sprinkler systems, before driving under them. Overhead guards must be used as protection against falling objects. It should be noted, however, that an overhead guard is intended to offer protection from the impact of small packages, boxes, bagged materials, etc., representative of the job application, but not to withstand the impact of a falling capacity load. Load backrest extensions must be used whenever necessary to minimize the possibility of the load or part of the load falling rearward.

When using a scoop or shovel attachment, such as on a front-end

A forklift is used to position this dock plate.

loader, operators should be told to work the front of the pile and not to undercut. Overhangs are dangerous.

Floors on which power trucks are used should be kept in good repair. They should have a load capacity great enough to support both truck and load. The combined weight of the heaviest truck and load should not exceed ⅕ the design strength of the floor. For example, a floor originally designed to bear 1,000 pounds per square foot (psf) should not be allowed to bear more than 200 psf.

The 5:1 ratio, although not extremely conservative, is quite important to establishing a low risk. A conservative safety factor provides an adequate margin of protection to offset possible miscalculations due to inadequate structural data, and misinformation about age, condition of floor members, type of floor, and other details.

Dock plates

Dock plates (bridge plates) should be fastened down in order to prevent them from "walking" or sliding when they are being used. Either the plate can be nailed down or holes can be drilled and pins or bolts dropped through them.

Plates should be large enough and strong enough to hold the

245

A bad accident may take place if dock plate is not secured or if a trailer is not properly blocked at a loading dock.

heaviest load, including the equipment. A factor of safety of 5 or 6 should be used. They can have their edges turned up, or have angle iron welded to the edges, in order to keep trucks from running over their edge.

Plates should be kept in good condition, with no curled corners or bends. They should be kept clean and dry—clear of snow and ice, and free of oil and grease. They are often protected from the weather by means of a canopy.

To use a plate, employees should slide it into place, not drop it. It must be kept under control and not allowed to bounce. Large plates are best set in place by mechanical equipment, or if this is not available, sufficient men should be assigned to the job. Handholds or other effective means must be provided on portable dock plates to permit safe handling.

The wheels of trucks or trailers backed up to the dock should be blocked or chocked to prevent movement when power trucks are run onto them for loading or unloading.

When not in use, dock plates should be placed safely to avoid tripping hazards. They should not be stored on edge so they can fall over.

Powered dockboards must meet the requirements of Commercial Standard CS 202–56 (1956), *Industrial Lifts and Hinged Loading Ramps*, published by the U.S. Department of Commerce.

For details of loading highway trucks and railroad cars, see these sections later in this chapter.

Conveyors

Do not let anyone ride on conveyors. There should be warning signs posted. If people must cross, crossover bridges should be provided to eliminate their having to crawl across the conveyor.

Gravity conveyors, usually the chute or roller type, should be equipped with warning devices, or protective devices to prevent people's hands from being caught in descending material or jammed between material and the receiving table. If the conveyor jams up, try first to free it from the top side. If necessary to enter the chute to free a jam, the person who enters should wear a life line and have another person stationed at the top of the chute to assist in case of emergency. People should be made aware of these hazards and trained to cope with them safely.

Pinch points—gears, chain drives, and revolving shafts—on conveyors should be guarded. People must leave these guards in place. Should the conveyor need servicing, the power should be shut off and the switch locked out.

Pneumatic or blower conveyors must be shut off and locked before inspections or repairs are made, in order to keep material from being blown into the worker's face or into the working area. If an inspection or service door must be opened while a system is operating, the inspector should wear goggles and, if necessary, respiratory equipment. Inspection ports, equipped with transparent coverings made of durable, nonflammable, shatter-resistant material, will allow inspection without stopping the operation.

Screw conveyors should be completely enclosed whenever possible. Their principal danger is that workmen may try to dislodge material or free a jam with their hands or feet, and then be caught in the conveyor. Screw conveyors should never be worked on unless the power is off and the switch locked out. Any exposed sections should be covered with a metal grate, with openings no larger than ½ in., or with a solid metal plate. The cover should be strong enough to withstand abuse. All removable covers should be interlocked so the power will be shut off when the covers are removed.

Cranes

Since August 27, 1971, all overhead and gantry cranes must meet the design specifications of American National Standard *Safety Code for Overhead and Gantry Cranes*, B30.2.0–1967 (OSHA regulation § 1910.179); and all crawler, locomotive, and truck cranes must meet American National Standard *Safety Code for Crawler, Locomotive, and Truck Cranes*, B30.5–1968 (OSHA regulation § 1910.180). Cranes installed before August 27, 1971, must be modified to conform to these requirements. (The equivalent requirements for construction are referenced in paragraphs 1926.550(e) and (b).)

Although many types of cranes are used in industry, safe operating procedures are pretty much the same.

Standard signals (illustrated) should be thoroughly understood by both operator and signalman. The supervisor should assign only one man as signalman for each crane, and he should tell the crane operator to obey only this man's instructions.

People working with or near a crane should keep out from under the load, should be alert at all times, and should watch warning signals closely. At a warning signal, they should move to a safe place immediately.

No crane should be loaded beyond its rated load capacity and the weight of all auxilliary handling devices such as hoist blocks, hooks, and slings must be considered to be part of the load rating. Substantial and durable rating charts with clearly legible letters and figures are to be fixed to the crane cab in a location easily visible to the operator while seated at his control station.

Hoist chain or hoist rope must be free from kinks or twists and must not be wrapped around the load, and loads must be attached to the load block hook by means of a sling or other approved device. Care must be taken to make certain that the sling clears all obstacles. The load must be well secured and properly balanced in the sling or lifting device before it is lifted more than a few inches. Before starting to hoist, make sure that multiple part lines are not twisted around each other.

Hooks should be brought over the load slowly to prevent swinging. During hoisting operation, see that there is no sudden acceleration or deceleration of the moving load and that the load does not contact any obstructions.

Cranes must not be used for side pulls, unless specifically authorized by a responsible person who can determine that the stability of

Standard two-hand signals are suitable for use with locomotive and other boom cranes.

the crane is not endangered and that the various parts of the crane will not be overstressed.

There must be no hoisting, lowering, or traveling while any

employee is on the load or hook, and operator must not carry loads over people. On overhead cranes, a warning signal must be sounded when starting the bridge and when the load or hook approaches near or over any person.

Brakes must be tested each time a load approaching the rated load is handled, and loads must not be lowered below the point where less than two full wraps of rope remain on the hoisting drum. When

Standard   one-hand   signals for overhead traveling and bridge cranes.

two or more cranes are used to lift a load, one qualified, responsible person must be in charge of the operation—he must analyze the operation and instruct all personnel involved in the operation as to the proper positioning, rigging of the load, and all movements to be made.

An operator must not leave his position at the controls while load is suspended. All necessary clothing and personal belongings must be stored in the cab, in such a manner as not to interfere with access or operation. Tools, oil cans, waste, extra fuses, and other necessary

250

Up she goes! Giant KC-97 tanker is raised by heavy cranes as practice for Air International Recovery System, which can lift and transport heavily damaged aircraft from airport runways in a matter of hours and without any secondary damage to the air frame.

articles must be kept in a toolbox and not allowed to lie loose in or about the cab. There must be a hand fire extinguisher kept in the cab. Access to the cab must be by fixed ladder, stairs, or platform, requiring no step over any gap exceeding 12 inches.

The upper load limit switch of each hoist must be tried out under no load conditions, at the beginning of each operator's shift. If the switch does not operate properly, an appointed person must be notified immediately. Hoist limit switch which controls the upper limit of travel of the load block must never be used as an operating control.

Inspections. Cranes, in regular service, must be inspected periodically, at intervals dependent upon the nature of the critical components of the crane and the degree of its exposure to wear, deterioration, or malfunction. Daily inspections must be made of all control mechanisms for maladjustment interfering with proper operation, including deterioration or leakage in air or hydraulic systems.

251

Frequent inspections, daily to monthly intervals or as specifically recommended by the manufacturer, (a) of all control mechanisms for excessive wear of components and contamination by lubricants or other foreign matter, as well as all safety devices for malfunction, (b) of crane hooks for deformations or cracks, for hooks with cracks or having more than 15 percent in excess of normal throat opening or more than 10-degree twist from the plane of the unbent hook; (c) for rope reeving not in compliance with manufacturer's recommendations; (d) of electrical apparatus for malfunctioning and excessive deterioration, dirt, and moisture accumulation.

Periodic inspections, from one- to twelve-month intervals or as specifically recommended by the manufacturer, cover the complete crane. These inspections include the items listed under frequent inspection, and in addition, include such items as (a) deformed, cracked, or corroded members in the crane structure and hook; (b) loose bolts or rivets; (c) cracked or worn sheaves and drums; (d) worn, cracked, or distorted parts, such as pins, bearings, shafts, gears, rollers, and locking devices; (e) excessive wear on brake and clutch system parts, linings, pawls, and ratchets; (f) load, boom angle, and other indicators over their full range, for any significant inaccuracies; (g) gasoline, diesel, electric, or other power plants for improper performance or noncompliance with safety requirements, excessive wear of chain-drive sprockets, and excessive chain stretch; and (h) travel steering, braking, and locking devices for malfunction and excessively worn or damaged tires. Determine if they constitute a safety hazard. Correct them. Records must be kept; see OSHA requirements.

Prior to initial use, all new and altered cranes must be inspected as mentioned earlier for periodic and frequent inspections. Cranes which are not in regular service and which have been idle for a period of one month or more, but less than six months, must be inspected in accordance with those as outlined for frequent inspections. Cranes which have been idle for a period of six months or more must be given a complete inspection in accordance with the requirements as called for in both frequent as well as periodic inspections, and standby cranes must be inspected at least semiannually in accordance with requirements for frequent inspections; cranes that were exposed to adverse environment must be inspected more frequently.

All ropes must be given a thorough inspection at least once a month, and a full written, dated, and signed report of rope condition must be kept on file where readily available to appointed personnel. Any deterioration, resulting in appreciable loss of original strength, such as reduction of rope diameter below nominal diameter, due to loss of core support, internal or external corrosion, or wear of outside wires, a number of broken outside wires and the degree of distribution or concentration of such broken wires, worn outside wires, corroded or broken wires at end connections, severe kinking, crushing, cutting, or unstranding, must be carefully noted and determination made as to whether further use of the rope would constitute a safety hazard.

All rope which has been idle for a period of a month or more due to shutdown or storage of a crane on which it is installed must be given a thorough inspection before it is placed in service. This inspection shall be for all types of deterioration and shall be performed by an appointed person whose approval shall be required for further use of the rope. A written and dated report of the rope condition shall be available for inspection.

On limited-travel ropes, heavy wear and/or broken wires may occur in sections that come in contact (*a*) with equalizer sheaves or other sheaves where rope travel is limited, or (*b*) with saddles. Particular care must be taken to inspect all ropes at these locations, furthermore particular care must be taken in the inspection of all non-rotating ropes.

Safety rules for electric overhead cranes are suggested by a crane manufacturer.

1. Only a regularly authorized operator should use a crane.
2. When on duty, he should remain in the crane cab ready for prompt service.
3. Before traveling, make certain the hook is high enough to clear all obstacles.
4. Under no circumstances may one crane bump another.
5. Examine the crane at every shift for loose or defective gears, keys, runways, railings, warning bells, signs, switches, sweep brushes, cables, and other parts. Report defects. Keep the crane clean and well lubricated.
6. No one should go to the top of a crane without first opening the main switch, placing a warning sign on it, and locking it out. He should unlock the switch and remove the sign promptly when he comes down.
7. After completion of a repair job, make sure that bolts, tools, and

When attaching or removing an electrically operated device attached to an overhead crane, make certain that the electrical switch in the cab has been opened. The crane operator should signal the hook-on man that this has been done.

other materials have been removed, so that damage to machinery will not result when the crane is started and so that nothing can fall off the crane. Keep tools, oil can, and other loose objects in a box provided for that purpose.

8. Do not carry a load over men on the floor. Sound a gong or siren when necessary. Men must not ride the load or the crane hook.

9. If power goes off, move controller to "off" position until power is again available.

10. See that the fire extinguisher on the crane is kept in good condition, and if used, that it is refilled.

11. Never let an operator run a crane if he is not physically fit to do so or if he is ill.

12. Do not drag slings, chains, or cable. After the load is taken off,

do not move the crane until the hook is lowered and the hook-on man has hooked the chain or cable.
13. Be sure the operator knows that if he is asked to move something he thinks is unsafe, that he check with his supervisor first.
14. Be sure that when an operator leaves the cage, that he leaves the main switch open. A magnetic crane must be empty and its controller turned off.
15. When an outside crane is parked at the end of a shift, the brake should be set or the crane chained to the track.
16. The operator should know that if his crane fails to respond correctly he should call the supervisor. Attempting to get out of difficulty by repeated operation may make the condition worse, not better.

Railroad cars

Spotting of railroad cars is usually done by switch engine crews. If it is necessary for the employees to move a railroad car, the supervisor should consider the following factors.

1. The facilities or equipment available to move the car.
2. The distance the car must be moved.
3. The number of cars to be moved.
4. Whether the track is level or on a grade.

After weighing these factors, the supervisor can decide whether to get power equipment, or move the car by hand, or take material to (or from) the car where it is. If at all possible, power equipment should be used to move the car.

If a car must be spotted or moved without power equipment, a car mover (which should be operated by only one man) can be used. If it takes two men to move a car, then each man should have a car mover and they should work on opposite wheels. All hand-actuated car movers should be equipped with knuckle guards. A stationary car puller, if available, should be used. When a car is being spotted or a full car is being moved to make room for an empty, these rules should be followed.

1. Check to make certain no one is working in the car or in adjacent cars to the one being moved.
2. Remove all bridge plates.

3. Move only one car at a time with hand equipment. If winches are used, do not overload them. The winch hook should be placed in the eye provided or the frame beneath the car. Do not hook to steps, ladders, or grab irons.

4. Make sure the car will not continue to roll after it has been moved far enough. (A man should be on the car ready to apply the brakes.)

5. Check for hazards to railroad trainmen. Particular attention should be given to proper warning signs, proper clearance from obstructions, and housekeeping along the tracks.

6. After a car has been moved to the proper location, set the brakes and block the wheels.

When moving a car, here are some precautions:

1. Never use a pinch bar or crowbar.

2. Never try to stop a car with a piece of lumber or metal on the track.

3. Never move a car toward the main line of the railroad if the slope is toward the main line. Derails must be left in derailing position.

4. Never move cars down an incline or sloping track, unless absolutely necessary.

5. Never use a truck or other road vehicle to push or pull a car.

6. Never leave a car mover in place under wheels. A car mover under pressure will fly out when a train crew moves the car.

When opening doors of boxcars, all persons should be alert to avoid having fingers caught, or freight fall out. A bar or ratchet hoist or puller should be used. Do not use trucks to open car doors as damaged doors may fall from runners.

Metal transfer plates must be securely anchored with bolts, and gangplanks and skids securely placed before using. Operators of mechanized equipment should avoid bumping doors or door posts.

Metal bands should be removed or cut to avoid dangling ends. Special equipment, such as DF bars, should be stacked in ends of car when empty and the car doors closed and latched before the car is to be moved.

Specially equipped cars with moveable bulkheads have instructions in the car which must be followed. When bulkhead partitions are difficult to move, examine the rollers to determine the reason for binding to avoid having the bulkhead fall on those working in the car.

Highway trucks

When loading or unloading highway trucks, be sure the brakes are set and wheel chocks placed under the rear wheels to prevent the trucks from rolling when they are boarded by powered industrial trucks. Use a fixed jack to support a semitrailer that is not coupled to a tractor in order to prevent up-ending of the semitrailer if a powered industrial truck operates in it. For more details of truck and truck terminal safety, see NSC's *Motor Fleet Safety Manual*.

Motorized equipment

Heavy-duty trucks, mobile cranes, tractors, bulldozers, and other motorized equipment used in the production of stone, ore, and similar materials and in construction work have been involved in frequent and often serious accidents. In general, prevention of these accidents requires:

1. Safe equipment.
2. Systematic maintenance and repairs.
3. Training operators in safe practices.
4. Training repairmen in safe practices.

Operation. Manufacturer's manuals contain detailed information on the operation of their equipment, and many driving practices are similar to those necessary for safe operation of highway vehicles. However, off-the-road driving involves special hazards requiring special training and safety measures. The modern heavy-duty truck or other off-the-road vehicle is a carefully engineered and expensive piece of equipment and warrants operation only by drivers who are qualified physically and mentally and by training and experience. For safety, these standards should be very high.

Even an experienced driver or operator should not be put to work until his knowledge and abilities have been determined. The time that is required for prospective drivers or operators to become thoroughly acquainted with the mechanical features of the equipment, safety rules, driver reports, and emergency conditions varies. In no case, should even an experienced man be permitted to operate equipment until the instructor or supervisor is satisfied with his abilities.

After a man has been trained, supervision has the important responsibility of seeing that he continues to operate in the way in which he was instructed.

Equipment. Safe operation of heavy equipment begins with the purchase of the machine. A good policy is to specify safeguards over exposed gears, safe oiling devices, hand holds, and other safety devices when the order is placed with the manufacturer. In any case, before equipment is put into operation, it should be thoroughly inspected and necessary safety devices installed and checked.

Both operators and repairmen, whether experienced or not, should be instructed in the recommendations of the manufacturer pertaining to lubrication, adjustments, repairs, and operating practices, and should be required to observe them. A preventive maintenance program is essential for safety and efficiency. Frequent and regular inspections and prompt repairs are the basis for effective preventive maintenance.

Generally, the operator is responsible for inspecting the mechanical conditions, such as hold-down bolts, brakes, clutches, clamps, hooks, and similar vital parts. Wire ropes should be kept lubricated (see manufacturer's instructions); for inspection, see page 263.

Many of the basic safety measures recommended for trucks also apply to motor graders and other types of earth-moving equipment. All machines should be inspected regularly by the operator, who also should promptly report any defects. The safety and efficiency of equipment are increased by scheduled maintenance.

Ropes, Chains, and Slings

There are special safety precautions that apply to using ropes, rope slings, wire rope, chains and chain slings, and storing chains. The supervisor should know the properties of the various types used and the precautions both in use and maintenance.

Fiber ropes

Fiber rope is used extensively in handling and moving materials. The rope is generally made from manila (abaca), sisal, hemp, or nylon. Manila or nylon ropes give the best uniform strength and service. Other types of rope on the market today include those made from polyester or polypropylene, which are adaptable to special uses.

Sisal and hemp are not as satisfactory as manila because their strength varies in different grades. Sisal rope is about 67 percent as strong as manila; hemp, about 50 percent as strong, but more resistant to atmospheric deterioration.

Manila rope is yellowish with a somewhat silvery or pearly luster. Sisal rope is also yellowish, but often has a green tinge and lacks the luster of manila. Its fibers tend to splinter and it is rather stiff.

The safe working loads and breaking strengths of manila, sisal, nylon, polyester, and polypropylene ropes are shown in the table.

Maintenance tips. Precautions should be taken to keep rope in good condition. Kinking, for example, strains the rope and may overstress the fibers. It may be difficult to detect a weak spot made by a kink. To prevent a new rope from kinking while it is being uncoiled, first lay the rope coil on the floor with the bottom end down. Then pull the bottom end up through the coil and unwind the rope counterclockwise. If it uncoils in the other direction, turn the coil of rope over and pull the end out on the other side.

Rope should not be stored unless it has been cleaned. Dirty rope can be hung in loops over a bar or beam and then sprayed with water to remove the dirt. The spray should not be so powerful that it forces the dirt into the fibers. After washing, the rope should be allowed to dry, and then be shaken to remove any remaining dirt.

Rope must be thoroughly dried out after it becomes wet, otherwise it will deteriorate quickly. A wet rope should be hung up or laid in a loose coil in a dry place until thoroughly dry. Rope will deteriorate more rapidly if it is alternately wet and dry than it will if it remains wet. Wet rope should not be allowed to freeze.

Rope should be stored in a dry place where air circulates freely about it. Air should not be extremely dry, however. Small ropes can be hung up, and larger ropes can be laid on gratings so air can get underneath and around them.

Rope should not be stored or used in an atmosphere containing acid or acid fumes, as it will quickly deteriorate. Signs of deterioration from this cause are dark brown or black spots on the rope.

Sharp bends should be avoided whenever possible because they cause extreme tension in the fibers. When cinching or tying a rope, be sure the object has a large enough diameter to prevent the rope from bending sharply. A pad can be placed around sharp corners.

If at all possible, a rope should not be dragged as this abrades

TABLE IX

TENSILE STRENGTH AND WEIGHT OF SYNTHETIC AND NATURAL FIBER ROPE

Size in Inches		Nylon		Polyester		Polypropylene		3-Strand Natural Fiber		
Diameter	Circumference	Pounds per 100 feet	Tensile Strength (lb)	Pounds per 100 feet	Tensile Strength (lb)	Pounds per 100 feet	Tensile Strength (lb)	Pounds per 100 feet	Tensile Strength (lb) Manila	Sisal
3/16	5/8	1.0	1,000	1.2	1,000	0.70	800	—	—	—
1/4	3/4	1.5	1,650	2.0	1,650	1.2	1,250	2.0	600	480
5/16	1	2.5	2,550	3.1	2,550	1.8	1,900	—	—	—
3/8	1-1/8	3.5	3,700	4.5	3,700	2.8	2,700	4.1	1,350	1,080
7/16	1-1/4	5.0	5,000	6.2	5,000	3.8	3,500	—	—	—
1/2	1-1/2	6.5	6,400	8.0	6,400	4.7	4,200	7.5	2,650	2,120
9/16	1-3/4	8.3	8,000	10.2	8,000	6.1	5,100	—	—	—
5/8	2	10.5	10,400	13.0	10,000	7.5	6,200	13.3	4,400	3,520
3/4	2-1/4	14.5	14,200	17.5	12,500	10.7	8,500	16.7	5,400	4,320
13/16	2-1/2	17.0	17,000	21.0	15,500	12.7	9,900	—	—	—
7/8	2-3/4	20.0	20,000	25.0	18,000	15.0	11,500	22.5	7,700	1,540
1	3	26.0	25,000	30.5	22,000	18.0	14,000	27.0	9,000	7,200
1-3/16	3-1/4	29.0	28,800	34.5	25,500	20.4	16,000	—	—	—
1-1/8	3-1/2	34.0	33,000	40.0	29,500	23.7	18,300	36.0	12,000	9,600
1-1/4	3-3/4	40.0	37,500	46.3	33,200	27.0	21,000	41.8	13,500	10,800
1-5/16	4	45.0	43,000	52.5	37,500	30.5	23,500	—	—	—

1-½	4-½	55.0	53,000	66.8	46,800	38.5	29,700	60.0	18,500	14,800
1-⅝	5	68.0	65,000	82.0	57,000	47.5	36,000	74.4	22,500	18,000
1-¾	5-½	83.0	78,000	98.0	67,800	57.0	43,000	89.5	26,500	21,200
2	6	95.0	92,000	118	80,000	69.0	52,000	108	31,000	24,800
2-⅛	6-½	109	106,000	135	92,000	80.0	61,000	125	36,000	28,800
2-¼	7	129	125,000	157	107,000	92.0	69,000	146	41,000	32,800
2-½	7-½	149	140,000	181	122,000	107	80,000	167	46,500	37,200
2-⅝	8	168	162,000	205	137,000	120	90,000	191	52,000	41,600
2-⅞	8-½	189	180,000	230	154,000	137	101,000	215	58,000	46,400
3	9	210	200,000	258	174,000	153	114,000	242	64,000	51,200
3-¼	10	263	250,000	318	210,000	190	137,000	299	77,000	61,600
3-½	11	316	300,000	384	254,000	232	162,000	—	—	—
3-⅝	11½	—	—	—	—	—	—	367	91,000	72,800
4	12	379	360,000	460	300,000	275	190,000	436	105,000	84,000

Recommended Factors of Safety: Nylon 9. Polyester 9. Polypropylene 6. Manila 5. Sisal 5.

These specifications of weight, circumference and strength for the synthetic ropes have been derived from tests made in accordance with the "Cordage Institute Standard Test Methods for Synthetic Fiber Ropes" dated November 10, 1960: revised December 8, 1965. They represent the Cordage Institute specifications (adopted January 13, 1966).

Tensile Strength figures are "Average." Minimum will be 5% below "Average." Pounds per 100 feet are "Average." Maximum weight 5% above these figures.

the outer fibers. If the rope picks up dirt and sand, abrasion within the lay of the rope will rapidly wear it out.

Lengths of rope should be joined by splicing. A properly made short splice will retain approximately 80 percent of the rope's strength. A knot will retain only about 50 percent. See details in Chapter 7, under Life lines, pages 190 and 191.

Rope should be inspected at least every 30 days, more often if it is used to support scaffolding on which men work. The following procedure will assist in the inspection of rope:

1. Check for broken fibers and abrasions on the outside.
2. Inspect fibers by untwisting the rope in several places. If the inner yarns are bright, clear, and unspotted, the strength of the rope has been preserved to a large degree.
3. Unwind a piece of yarn 8 in. long and ¼ in. diameter from the rope. Try to break this with your hands. If the yarn breaks with little effort, the rope is unsafe.
4. Inspect rope used around acid or caustic daily. If black, or rusty brown spots are noted, test the fibers as described in steps 1, 2, and 3, and discard all rope that fails the tests.
5. As a general rule, rope that has lost its pliability or stretch, or in which the fibers have lost their luster and appear dry and brittle, should be viewed with suspicion and replaced with a new rope. This is particularly important if the rope is used on scaffolding or for hoisting where, if the rope breaks, workers may be injured or property may be damaged.

Rope slings

Because of the high tensile strength needed for a sling, fiber rope slings should be made of manila only. Below are some precautions to be observed in using manila rope slings.

1. Make sure the sling is in good condition and of sufficient strength (take into account the factors of the splice and the leg angles of the sling). Hooks, rings, and other fittings must be properly spliced.
2. Reduce the load by one-half after the sling has been in use for 6 months, even though the sling shows no sign of wear.
3. If the sling shows evidence of cuts, excessive wear, or other damage, destroy it.

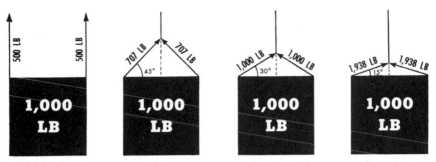

Decreasing the angle between the sling leg and horizontal increases the stress on each leg of the sling, even though the load stays the same.

Whether fiber rope or wire rope slings are used, certain general precautions should be taken. These were given under "Fiber ropes—Maintenance tips" listed earlier. In addition, it is best to keep the legs of slings as nearly vertical as possible. (See the drawing above).

Wire rope

Wire rope is used widely instead of fiber rope because it has greater strength for the same diameter and weight, its strength is constant when wet or dry, it has constant length under varying weather conditions, and it has greater durability.

Wire rope is made up according to its intended use. A large number of wires give flexibility. Fewer strands with fewer wires is less flexible.

A wire rope used for general hoisting should not be subjected to a working load greater than ⅕ its breaking strength—a factor of safety of 5. Greater factors of safety (6 or 7, for example) can always be used for extra security. Ropes should be thoroughly inspected at least once a month, and full written, dated, and signed reports kept on file. OSHA regulations spell out the details of this inspection.

Wire rope should be lubricated at regular intervals to prevent rust and excessive wear. Sheaves for rope should be as large as possible. The less flexible the rope, the larger the sheave or drum diameter must be, otherwise the rope will be bent too sharply. See the table. Rope should be wound in one layer only. Using several layers will mash or jam the rope and shorten its life.

Sheaves and drums should be aligned as much as possible to

TABLE X

TREAD DIAMETERS OF SHEAVES AND DRUMS

Rope Classification	Average Recommended	Minimum
6 x 7. . . . . .72 times rope diameter		42 times rope diameter
6 x 19. . . . . .45 " " "		30 " " "
6 x 37. . . . . .27 " " "		18 " " "
8 x 19. . . . . .31 " " "		21 " " "

prevent excessive wear of rope. Reverse bending of rope, in which rope is bent first in one direction and then in the other, should be avoided, as this wears out wire rope faster than anything else.

Fittings should be properly selected and attached if they are to be safe. Some of the principal wire rope attachments are:

> Babbitt- or zinc-coated connections
> Wedge sockets
> Swaged attachments
> Thimble-with-clip connections
> Three-bolt clamps
> Spliced eye-and-thimble connections
> Crosby clips

The crosby clip is probably the most commonly used fitting. It is important that these clips be installed properly. Wedge sockets are dependable and will develop approximately 75 to 90 percent of the strength of the rope. Swaged attachments are satisfactory for smaller diameter (¼- to 1-inch) ropes.

Wire rope should be spliced only by experienced men who have been trained to do it properly. Splices should be tested to twice the load they are expected to carry.

Chains and chain slings

Alloy steel chain, approximately twice as strong—size for size—as wrought iron chain, has become the standard material for chain slings.

One advantage is that it is suitable for high-temperature operations. Continuous operation at temperatures of 800 F (the highest temperature for which continuous operation is recommended) requires a reduction of 30 percent in the regular working load limit. For inter-

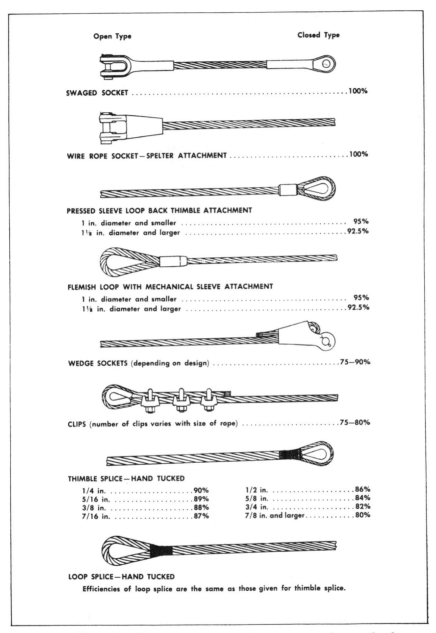

Open Type   Closed Type

SWAGED SOCKET 100%

WIRE ROPE SOCKET—SPELTER ATTACHMENT 100%

PRESSED SLEEVE LOOP BACK THIMBLE ATTACHMENT
  1 in. diameter and smaller 95%
  1⅛ in. diameter and larger 92.5%

FLEMISH LOOP WITH MECHANICAL SLEEVE ATTACHMENT
  1 in. diameter and smaller 95%
  1⅛ in. diameter and larger 92.5%

WEDGE SOCKETS (depending on design) ....75—90%

CLIPS (number of clips varies with size of rope) ....75—80%

THIMBLE SPLICE—HAND TUCKED
  1/4 in. ....90%   1/2 in. 86%
  5/16 in. ....89%   5/8 in. 84%
  3/8 in. ....88%   3/4 in. 82%
  7/16 in. ....87%   7/8 in. and larger...........80%

LOOP SPLICE—HAND TUCKED
  Efficiencies of loop splice are the same as those given for thimble splice.

Efficiencies of attaching fittings to wire rope, in percentage of strength of rope.

265

mittent service, these chains may be used at temperatures as high as 1,000 F, but at only 50 percent of the regular working load limit.

An alloy steel chain should never be annealed when the chain is being serviced or repaired because the process reduces the hardness of the steel, and this, in turn, reduces the strength of the chain. Wrought iron chain, on the other hand, should be annealed periodically

TABLE XI

WORKING LOAD LIMITS AND BREAK TEST LIMITS
FOR ALLOY STEEL CHAIN*

Nominal Size of Chain Bar in Inches	Working Load Limits in Pounds	Minimum Break Test in Pounds
¼	3,250	10,000
⅜	6,600	19,000
½	11,250	32,500
⅝	16,500	50,000
¾	23,000	69,500
⅞	28,750	93,500
1	38,750	122,000
1⅛	44,500	143,000
1¼	57,500	180,000
1⅜	67,000	207,000
1½	79,500	244,000
1¾	94,000	325,000

*Source: *Specification for Alloy Steel Chain*, American Society for Testing and Materials, A-391-58. The strengths for wrought iron chain is slightly less than half these figures.

by the manufacturer (or persons specially trained in the operation) if it is used where its failure might endanger human life or property.

Impact loads, caused by faulty hitches, bumpy crane tracks, and slipping hook-ups, can materially add to the stress in the chain. The impact resistance of heat-treated alloy steel chain does not increase in proportion to the strength of the chain. Under a full working load, it will fail under impact before fully loaded wrought iron, or heat-treated carbon steel chain will.

Chain slings should preferably be purchased complete from the manufacturer, and whenever they need repair, they should be returned to their manufacturer. The manufacturer should also be queried about details of strength and specifications of the slings.

Load and wear. Chains should not be overloaded—the working load limit and break test limit for alloy steel chain is given in the table.

TABLE XII

CORRECTION TABLE FOR THE REDUCTION
OF WORKING LOAD LIMITS OF CHAIN DUE TO WEAR*

	REDUCE WORKING LOAD LIMITS OF CHAIN BY THE FOLLOWING PER CENT WHEN DIAMETER OF STOCK AT WORN SECTION IS AS FOLLOWS:		
NOMINAL ORIGINAL CHAIN SIZE, INCHES	*5 Per Cent*	*10 Per Cent*	*Remove from Service*
¼ 0.250........	0.244	0.237	0.233
⅜ 0.375........	0.366	0.356	0.335
½ 0.500........	0.487	0.474	0.448
⅝ 0.625........	0.609	0.593	0.559
¾ 0.750........	0.731	0.711	0.671
⅞ 0.875........	0.853	0.830	0.783
1 1.000........	0.975	0.949	0.895
1⅛ 1.125........	1.100	1.070	1.010
1¼ 1.250........	1.220	1.190	1.120
1⅜ 1.375........	1.340	1.310	1.230
1½ 1.500........	1.460	1.430	1.340
1⅝ 1.625........	1.590	1.540	1.450
1¾ 1.750........	1.710	1.660	1.570
1⅞ 1.875........	1.830	1.780	1.680
2 2.000........	1.950	1.900	1.790

*From American National Standard B30.2, *Safety Code for Cranes, Derricks, and Hoists.*

Some of the principal causes of chain failure are:

1. Embrittlement caused by cold working of the metal surface.
2. Failure of the weld.
3. Repeated severe bending or deformation of the links.
4. Metal fatigue.
5. Brittleness caused by defects in the metal.
6. Tensile failure—full elongation of the links.

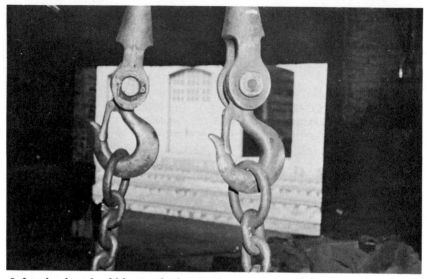

Safety latches should be standard on hooks used on chain slings. Hooks should be made of forged or laminated steel.

How much wear a chain can stand is determined by its use, the original factor of safety, and the loads the chain is expected to carry. A regular schedule of checking and calipering the links should be set up for each chain, depending on usage. To check elongation, chain should be calipered in gage lengths of 1 to 3 (preferably 5 links when new),* and that record should be kept. The chain (or sling) should be discarded if it shows a stretch (includes link wear) of more than 5 percent. Some states specify a chain must be destroyed or discarded when any 3-foot length is found to have been stretched ⅛ of a link-length.

A broken chain should never be spliced by inserting a bolt between two links. Neither should one link be passed through another and a nail or bolt inserted to hold them.

Use. Before making a lift, the men should make sure that there are no kinks, knots, or twists in the chain. They should lift the load gradually and uniformly and make sure it is not merely attached to the tip of the hook. They should never hammer a link over the hook as this will also stretch the hook and stretch the links of the chain.

* Section 2813 of American National Standard B30.2, *Safety Code for Cranes, Derricks, and Hoists.* See "Useful References."

When lowering a crane load, brakes should be applied gradually to bring the load to a smooth stop.

Each chain should be tagged to indicate its load capacity, date of last inspection, date of purchase, and the type of material from which the chain is made. This information must not be stamped on the lifting links, however, as it may set up stress points that weaken the chain.

Chain attachments (rings, shackles, couplings, and end links) should be made of the same material to which they are fastened. Hooks should be made of forged steel or laminated steel, and should be equipped with safety latches. Replace hooks that have been over-loaded, or loaded on the tips, and have a permanent set greater than 15 percent of the normal throat opening.

Storing chains

Proper chain storage serves two purposes: preserving the chain and promoting good housekeeping. Chain is best stored on racks inside dry buildings that have a fairly constant temperature. Chain should be stored where they will not be run over by trucks or mobile equipment, and where they will not be exposed to corrosive chemicals.

Storage racks and bins should be constructed so that air will circulate around the chain and keep it dry. Power equipment should be used to lift large and heavy chain. Attempting to lift a chain manually may injure a person.

Materials Storage

Both temporary and permanent storage must be neat and orderly. Materials piled haphazardly or strewn about increase the possibility of accidents to employees and of damage to materials.

The warehouse supervisor must direct the storage of raw materials (and sometimes processed stock) that are kept in quantity lots for some time. The production supervisor is usually responsible for storage of limited amounts of materials and stock for short periods close to the processing operations.

Planned materials storage minimizes the handling necessary to bring materials into production, as well as remove finished products from production to shipping.

Tin plate strip stock is heavy and sharp. Full loads can be banded to keep strips together, but when bands are broken, it is these wooden stakes that keeps a partial load from falling. Even with these precautions, it still takes supervision to make sure that the operator, handler, trucker, and anyone else who works with the load keeps the stakes in place.

When supervisors plan materials storage, they should make sure that materials do not obstruct fire alarm boxes, sprinkler system controls, fire extinguishers, first aid equipment, lights, and electric switches and fuse boxes. All exits and aisles must be kept clear at all times.

There should be at least 18 in. clearance below sprinkler heads to reduce interference with water distribution. This clearance should be increased to 36 in. if the material being stored is very flammable.*

Aisles that carry one-way traffic should not be less than 3-ft wider than the widest vehicle is when it is loaded. If materials are to be handled from the aisles, the turning radius of the power truck may also need to be considered. Employees should be told to keep ma-

---

* OSHA regulations §1910.159(e). Also see NFPA Standard No. 13, *Installation of Sprinkler Systems.* See page 387, later in this Manual.

270

terials out of the aisles and out of loading and unloading areas, which should be marked with white painted lines.

Storage is facilitated and hazards are reduced by use of bins and racks. Material stored on racks, pallets, or skids can be moved easily and quickly from one work station to another with less material damage and fewer employee injuries. When possible, material piled on skids or pallets should be cross-tied.

In an area where the same type of material is stored continuously, it is a good idea to paint a horizontal line on the wall to indicate the maximum height to which material may be piled. This will help keep the floor load within the proper limits and the sprinkler heads in the clear.

Containers and other objects

Because packing containers vary considerably in size and shape, height limitations for stacking these materials vary. The weight of the container itself must also be figured when keeping within the proper floor-loading limit.

Sheets of heavy wrapping paper placed between layers of cartons will help prevent the pile from shifting. Better still, cross-tying will prevent shifting and sagging and permit higher stacking. Wire-bound containers should be placed so that any sharp ends do not stick into the passageway. When stacking corrugated paper cartons on each other, remember that increased humidity may cause them to slump.

Bagged materials can be cross-tied with the mouths of the bags toward the inside of the pile, so contents will not spill if the closure breaks. Piles over 5-ft high should be stepped back one row. Step back an additional row for each additional 3 ft of height. Sacks should be removed from the top of the pile, not half way up or from a corner. Sacks should be removed from an entire layer before the next layer is started. This keeps the pile from collapsing, possibly damaging material and injuring people.

Small diameter bar stock and pipe are usually stored in special racks, so located that when stock is removed, passers-by are not endangered. The front of the racks should not face the main aisle, and material should not protrude into the aisles. When the storage area is set up, the floor load which this type of storage entails should be considered.

271

Bar stock and large diameter pipe should be piled in layers separated by strips of wood or by iron bars. Storage should not extend into the aisles.

Larger diameter pipe and bar stock should be piled in layers separated by strips of wood or by iron bars. If wood is used, it should have blocks at both ends; if iron bars are used, the ends should be turned up. Again the floor load must be considered.

Lumber and pipe have a tendency to roll or slide. When men are to make a pyramid pile, they should be told, therefore, (*a*) to set the pieces down rather than to drop them, and (*b*) not to use their hands or feet to stop rolling or sliding material. These practices could cause serious injury.

Sheet metal usually has sharp edges. It should be handled with hand leathers or leather gloves, or gloves with metal inserts. Large amounts should be handled in bundles by power equipment. These should be separated by strips of wood to facilitate handling when the material is needed for production and to lessen chances of shifting or sliding of the piles of material.

Tin plate strip stock is heavy and razor sharp. Should a load, or partial load, fall, it could badly injure anyone in the way. Two measures can be taken to prevent spillage and injuries: (*a*) band the stock after shearing, and (*b*) use wooden or metal stakes around the stock tables and pallets that hold the loads. It is the responsibility of the supervisor and all who handle the bundles to make sure the load is banded properly and that the stakes are in place when the load is on the table.

Packing materials, such as straw, excelsior, or shredded paper, should be kept in a fire-resistant room, provided with sprinklers and dust-proof electric equipment. Materials received in bales should be kept baled until they are used, then only enough material for immediate use, or for one day's supply, should be taken to the packing room and placed in metal (or metal-lined wood) bins, the covers of which should have fusible links so they will close automatically in case of fire. To prevent injury in case a counterweight rope should break, the weight should be boxed in.

Barrels or kegs stored on their sides should be pyramided, with the bottom rows securely blocked to prevent rolling. If piled on end, the layers should be separated by planks. If pallets are used, they should be large enough to prevent overhang.

Hazardous liquids

Acid carboys are best handled with special equipment, such as carboy trucks. Boxed carboys should generally be stacked not higher than two tiers, never higher than three. Not more than two tiers should be used for carboys of strong oxidizing agents, such as concentrated nitric acid or concentrated hydrogen peroxide. The best method of storage would be in specially designed racks.

Before anyone handles carboy boxes, he should inspect them thoroughly to make sure that nails have not rusted out or that the wood has not been weakened by acid action. Before being shipped back to the supplier, empty carboys should be completely drained and their stoppers replaced.

The safest way to draw off liquid from a carboy is to use suction from a vacuum pump, or to use a syphon started by either a rubber bulb or an ejector. Another method is to use a carboy inclinator of the type that holds the carboy at the top, sides, and bottom, and returns the carboy to an upright position automatically when it is released. Pouring by hand or starting pipettes or syphons by mouth suction to draw off the contents of a carboy should never be permitted.

When liquids are handled in portable containers, contents should be clearly labeled. Substances should be stored separately in a well-lit area so that people will not make mistakes when drawing liquids out.

273

Storage rack reduces hazards, aids identification, and permits easy grounding of each drum. Note grates on the drip pans in front of the storage racks. The whole storage area has explosion-proof electrical equipment.

Your purchasing agent should specify that all chemicals are properly labeled showing contents, hazards, and precautions. Labels were discussed in Chapter 6, Industrial Hygiene.

Drums containing hazardous liquids should be placed in a rack, not stacked. It is best to have a separate rack for each type material being handled. Racks permit good housekeeping, easy access to the drums for inspection, and easy handling. Racks permit drums or barrels to be emptied through the bung. Self-closing spigots should be used, particularly where individuals are allowed to draw off their own supplies. If there is a chance that spigots might be hit by material or equipment, drums should be stored on end and a pump used to withdraw the contents.

Drums containing flammable liquid, and the racks that hold them, should be grounded, and the smaller container into which the drum is being emptied should be bonded to the chime of the drum by a flexible wire with a battery clip or C-clamp at each end.

Storage areas for liquid chemicals should be well ventilated. Natural ventilation is better than mechanical, because the system may

274

Acid carboys are best handled with special equipment, such as this carboy truck.

fail. The floor should be made either of concrete or of some other resistant material treated to reduce absorption of the liquids. The floor should be pitched toward one or more drains that are corrosion resistant and easily cleaned.

Where caustics or acids are stored, handled, or used emergency flood showers or eyewash fountains should be available. Men should be provided with chemical goggles, rubber aprons, boots, and gloves, and other protective equipment necessary to handle the particular liquid. (See Chapter 7, Personal Protective Equipment.)

Each type of hazardous material requires special handling techniques or special protective clothing. In all cases, the supervisor or the purchasing agent should consult the safety professional or the safety department.

Additional precautions to take when handling hazardous liquids:

1. Wear proper protective equipment.
2. Keep floors clean. Do not allow them to become slippery.
3. Drain siphons, ejectors, and other emptying devices completely before removing them from carboys.

4. To dilute acid, always add it to water, never add water to acid. Add the acid slowly and stir constantly with a glass implement.
5. In case of an accident, give first aid: flush burned areas with lots of water. Call a doctor immediately.

Combination eyewash fountain and emergency shower. Fountain is pedal-actuated so that both hands can be free to open eyelids.

6. Do not force hazardous chemicals from containers by injecting compressed air. The container may burst or the contents may ignite.
7. Do not try to wash or clean a container unless its label specifically requires that it be cleaned before it is returned.
8. Do not store hazardous chemicals in glass or other containers near heat or steam pipes, or where strong sunlight will strike them. The contents may expand and may cause a fire or explosion. Bottles may concentrate sunlight to ignite a fire in nearby combustibles.

9. Do not stir acids with metal implements.
10. Do not store chemicals or solvents in dark or poorly lit areas. The wrong chemical may be selected.
11. Do not pour corrosives by hand from a carboy.
12. Do not move a carboy unless it is securely stoppered and wired.

Pipelines to outside storage tanks are safer if large quantities of liquid chemicals or solvents are used, because they reduce and localize any spillage.

When hazardous liquids are used in small quantities, only enough for one shift should be kept on the job. In many cases, local ordinances limit the amount of hazardous materials which can be stored or processed inside a building. The main supply should be stored in tanks located in an isolated place.

Pipelines should be color coded to identify their contents. Basic colors for identification of contents of pipes by class and methods of providing specific identification were discussed in Chapter 8, Industrial Housekeeping. Should outside contractors work in the area, they should be adequately informed of piping systems and any hazards.

When a valve is to be worked on, it should be closed and the section of pipe in which it is located should be thoroughly drained before the flanges are loosened.

As a temporary measure, the opening between faces of a flange may be covered with a piece of sheet lead, and this wrapped around to make a sleeve while the flange bolts are being loosened and the faces separated. If the flanges do not separate readily, a strong nail can be driven through the lead shield into the flange joint to part it.

Workmen should loosen the flange bolts farthest away first so that the pipe will drain away from them. A blind (or blank) should be inserted between the flanges as soon as they are parted. If a line is opened often, a two-ended metal blank can be permanently installed on a flange bolt so that the blank can be pivoted on the bolt when the flanges are parted. One end of the blank is shaped in the form of a gasket, to permit flow of liquid. The other end is blind and is swung to cover the pipe opening when the flanges are unbolted and parted.

Tank cars should be protected on sidings by derails and by blue stop flags or blue lights before they are loaded or unloaded. Hand brakes should be set and wheels chocked. Before the car is opened, it should be bonded to the loading line. The track and the loading or unloading rack should be grounded and all connections checked regularly.

Chemical tank cars should be unloaded through the dome rather than through the bottom connection.

Rack is designed for safe storage of compressed gas cylinders.

Gas cylinders

Compressed gas cylinders should be stored on end on a smoother floor. All cylinders should be chained or otherwise fastened firmly against a wall, post, or other solid object. Different kinds of gases should be either separated by aisles or stored in separate sections of the building or storage yard. Empty cylinders should be stored apart from full cylinders.*

To prevent cylinders from being run against or knocked over, keep storage areas away from heavy traffic. Do not store cylinders of flammable gases near highly flammable substances. During winter,

* More details in OSHA regulations §1910.110 and *Liquefied Petroleum Gases, Storage and Handling,* Standard No. 58, published by the National Fire Protection Association. Storage of welding gases is detailed in OSHA regulations §1910.252 (a)(2). Compressed gases are covered in OSHA regulations §1910.101 and in CGA Pamphlet P–1–1965.

let cylinders that become caked with ice and snow thaw out gradually inside before opening them. Do not use a steam hose, or the heat may melt the fusible plug and release the cylinder contents. Do not use cylinders as rollers for heavy equipment.

Handle all cylinders with care—a cylinder marked "empty" just might not be.

To transport cylinders, use a carrier that does not allow excessive movement, sudden and violent contacts, and upsets. When a two-wheeled truck with rounded back is used, chain the cylinder upright. Never use a magnet to lift a cylinder. For short-distance moving, a cylinder may be rolled on its bottom edge, but never dragged. Cylinders should never be dropped or permitted to strike one another. Protective caps must be kept on cylinders when cylinders are not being used.

When in doubt as to how to handle a compressed gas cylinder, or how to control a particular type of gas once it is released from the cylinder or other source, ask the safetyman or the safety department.

Combustible solids

Bulk storage of grains and granular or powdered chemicals or other materials presents fire and explosion hazards. Many materials which are not considered hazardous in solid form often become quite combustible when finely divided. Some of these are carbon, fertilizers, food products and by-products, metal powders, resins, waxes and soaps, spices, drugs, and insecticides, wood, paper, tanning materials, chemical products, hard rubber, sulfur, starch, and tobacco. Though it is long, this list is by no means complete—it only shows the general categories of explosive dusts.

To avoid dust explosions, prevent formation of an explosive mixture or eliminate all sources of ignition. Either the dust must be kept down, or enough air supplied to keep the mixture below the combustible limit.

Good housekeeping and good dust-collection equipment will go far toward preventing disaster. Formation of layers of dust on floors and on other structural parts of a building should be prevented. A dust explosion is a series of explosions, or a progression of explosions. The initial explosion is very small, but it shakes up dust that has collected on rafters and equipment so that a second explosion is created.

This progress usually continues until the result is the same as if a large explosion took place.

Wearing protective equipment is important when dust is toxic. Such equipment may range from a respirator to a complete set of protective clothing. The National Safety Council, a state or local safety council, or a state or local governmental health and safety agency can recommend the proper type of protective clothing for specific jobs. See the recommendations in Chapter 7.

Bins in which solids are stored should have a sloping bottom to allow material to run out freely and to prevent arching. For some materials, especially if they are in-process, the bin should have a vibrator or agitator in the bottom to keep the material flowing.

If larger rectangular bins, open at one side, are entered with power shovels or other power equipment, men should take care not to undercut the material, and thereby endanger themselves if arched material should suddenly give way. Where practical, tops of bins should be covered with a 2-in. mesh screen, or at least by a 6-in. grating (or parallel bars on 6-in. centers) to keep people from falling into the bins.

Tests for oxygen content and for presence of toxic materials should be made before anyone enters a bin or tank for any purpose. If such tests indicate the need for protection, men should don air supplied masks, in addition to a safety belt and life line, before entering. Filling equipment should be made inoperative so that it cannot be started again except by the workman after he leaves the bin or by his immediate supervisor after he has checked the man out of the bin. When a person is working in a bin, he should have a companion employee, equipped to act in case of emergency, stationed on top of the bin. Bins should be entered from the top only.

If flammable vapors, dusts, or other flammable air contaminants are found to be present, workers should use only spark-resistant equipment and electrical equipment that is approved by Underwriters' Laboratories, Inc., for such hazardous locations.

CHAPTER 10

Guarding Machines and Mechanisms

Every supervisor recognizes he is responsible for directing people in producing a product or in providing a service at a profit for his firm. When mechanical equipment or production machinery is involved, he is responsible for keeping it running efficiently.

- When equipment is shut down for any unscheduled reason, production capacity and cost of operation are directly affected.

- When an employee is injured while operating unguarded machines or equipment, production capacity and cost of operation are again affected.

Nearly one-fifth of all permanent partial disabilities result from injuries associated with machinery.* Poorly designed, improperly safeguarded, or unguarded machinery or equipment is a positive threat to production capacity and to the well-being of employees.

An injury can take an operator away from his job for a long time, even permanently. This person must be replaced if the department is to keep going. Even if a qualified person is transferred from another job, somewhere down the line a new person must eventually be hired. Finding the right person takes time. Hiring him takes time. Training

* NSC *Accident Facts*, 1972 ed.

A screen encloses the V-belt drive for this pipeline pumping unit from its 200-hp electric motor.

takes time. Both the time and the delay cost money.

Where does safeguarding fit into this picture? Because no matter how much training a man may get, no matter how much experience he may have, he can still be injured by a hazardous, unguarded machine. Even the most skilled professional cannot always be 100 percent alert. There is no method yet devised that will keep a person's mind on his work every minute of the working day.

What about the deceptive machine—the one that appears so easy to run that the supervisor allows semi-skilled, or even unskilled persons to operate it? The one that is so "simple" that the supervisor does not even train the person? Often the results are drastic—in personal injury, machine damage, and material damage.

Safeguarding reduces the possibility of "human error," mechanical failure, poor design, and electrical failure. It is vital that the supervisor have a good understanding of the principles of how safeguarding can help him.

Danger zones of this knitting machine are safeguarded.

Principles of Guarding

"Guarding" is frequently thought of as being concerned only with the point of operation or with the means of power transmission. Although guarding these hazards is required, guarding can also prevent injuries from other causes, both on and around machines and from equipment and damaged material.

Guards or barriers can protect against, and prevent injuries from these sources:

1. Direct contact with exposed moving parts of a machine—either points of operation on production machines (such as power presses, machine tools, or woodworking equipment), or power-transmitting parts of mechanisms (such as gears, pulleys and sheaves, slides, or couplings).
2. Work in process, such as pieces of wood that kick back from a power

Safety cables (arrows) should be attached to machinery to hold in place all parts that probably can fail.

ripsaw, or metal chips that fly from tools or from abrasive wheels.
3. Machine failure, which usually results from lack of preventive maintenance, overloading, metal fatigue, or abuse.
4. Electrical failure, which may cause malfunctioning of the machine, or cause electrical shock or burn.
5. Operator error or human failure caused by distraction, worry, zeal, anxiety, misunderstanding, indolence, deliberate chance-taking, anger, illness, fatigue, and so on.

Positive prevention of injury-producing accidents on machinery can be assured through the installation of safeguards or through engineering revision and redesign. Injury-producing accidents are inevitable where equipment with dangerous moving parts is operated without guards or with incomplete or ineffective guards.

284

The experience of more than six decades of organized accident prevention has demonstrated that it is unwise to rely entirely on education and training of the operator or upon operator cooperation. Many factors affect a person's attitude and judgment, and even his ability to concentrate. Even a skilled person who is disturbed by mental, emotional, or physical problems cannot pay strict attention to his production responsibilities and cannot give his best efforts.

Benefits of good (safe)guarding

Possibly the prime benefit of good safeguarding, in the eyes of the supervisor, is that in many instances it improves production. As long as the operator is afraid of the machine or is afraid of getting close to moving parts, he cannot pay strict attention to his production responsibilities. Once his fear has been removed, a worker can concentrate on the operation at hand and usually produce more and better products.

Well designed and carefully maintained safeguards assure workers that management means what it says about its sincere desire to prevent accidents. When an employee sees that his employer is really interested, he will be more inclined to contribute to the safety effort by working safely.

Whenever moving parts are left exposed, objects may fall into them or get caught in them and damage the machine or material in process. Such damage may cause expensive shutdowns and loss of raw material, or finished product.

A good safeguarding program can mean much in terms of employee morale, especially when the men who will have to use the safeguards are consulted before the safeguards are made or bought. Often workers have ideas which contribute to both safety and economy. Even if they have no suggestions, they feel flattered at having been consulted. They now take a personal interest in the project, and are, therefore, less likely to remove the guards or barriers or fail to replace them. Finally, a good safeguarding program will help reduce violation litigations and help comply with state and federal regulations.

The first part of this chapter will discuss guarding of points of operation and of power-transmitting parts of machinery or equipment. Part II will discuss safeguarding of mechanisms. Part III will cover types of guards and their maintenance. For these discussions, a list of definitions will be helpful.

Definitions

These definitions will clarify the terms used in this chapter.

- *Danger zone* (*pinch point*). A point at which a person can get caught between (*a*) the moving parts of a machine, (*b*) moving and stationary parts, or (*c*) material and moving parts. It does not describe hazards caused by tooling at the point of operation of a power press.
- *Exposed to contact.* An object or moving part is located in a place where part of a person's body can come into contact with it and be injured.
- *Device* (*point of operation*). A term used in metal stamping operations to designate a machine control or attachment that allows the operator access to the point of operation for loading or unloading dies. It (*a*) restrains the operator from inadvertently reaching into the point of operation, (*b*) prevents normal operation until the operator's hands are removed, or (*c*) pulls the operator's hands from the point of operation as the dies close.
- *Guard.* A barrier that prevents entry of the operator's hands or fingers into the point of operation. ANSI Std. B11.1–1971, *Safety Requirements for Construction, Care, and Use of Mechanical Power Presses,* reserves the use of the word "guard" exclusively for referring to barriers designed for safeguarding at the point of operation. Therefore, the word "safeguard" is used for a barrier or cover that protects other danger zones. "(Safe)guard" refers to either.
- *Interlock.* A device that interacts with another device to govern succeeding operations in the same or an allied machine or equipment. See Part III.
- *Pinch point.* See "Danger zone."
- *Point of operation.* That part of a machine where stock is actually placed and kept during blanking, shaping, or other operation, including such other parts as may offer a hazard to the operator's inserting or manipulating stock or material.
- *Prime mover.* An engine used as a source of power. It may be steam, gas, oil, or air operated. Electric motors and steam or hydraulic turbines are included.
- *Securely fastened.* When something is so well secured that it cannot be moved or removed under normal or reasonably foreseeable conditions.

Part I—(Safe)guard Design

It is easier to establish effective methods of safeguarding power transmissions than it is for guarding points of operation, because power transmissions are far more standardized. The American National Standard *Safety Code for Mechanical Power-Transmission Apparatus*, B15.1, applies to all moving parts of mechanical equipment, excluding the point of operation (which is covered by B11.1). It illustrates and recommends materials, specifications, and extent of coverage needed for best protection. Some of the tables in this chapter are taken from this B15.1 Standard.

There are also ANSI Standards for other types of machinery—power presses, forging machines, woodworking machines, paper mill equipment, rubber mills, and calenders for example.

Most states and OSHA regulate the (safe)guarding of mechanical equipment. Occupational Safety and Health Standards, Part 1910, Subpart "O," cover safeguarding of machinery and equipment. It is a violation to ignore these regulations.

Whether (safe)guards be built by a user or purchased from a manufacturer, they should meet certain performance or design standards, or both, if they are to be effective. If a company does not have a safety engineer, it should be able to get the advice of an insurance company safety engineer, state factory inspector, or local safety council engineer as to what design would be most effective to safeguard a specific operation. The National Safety Council offers guidance to members. The NSC *Accident Prevention Manual* and *Guards Illustrated* give many ideas in addition to those given here.

To be generally acceptable, a (safe)guard should:

1. Conform to applicable American National Standards and to the requirements of OSHA and/or the state inspection department having jurisdiction.
2. Be considered a permanent part of the machine or equipment.
3. Afford maximum positive protection, not only for the operator, but also for passers-by.
4. Prevent access to the danger zone or point of operation during operation.
5. Not weaken the structure of the machine.
6. Be convenient. It must not interfere with efficient operation of the

TABLE XIII

STANDARD MATERIALS AND DIMENSIONS FOR MACHINERY SAFEGUARDS

Size of Filler Materials

Material	Clearance from Moving Part at All Points (inches)	Largest Mesh or Opening Allowable B (inches)	Minimum Gauge (U.S. Standard) or Thickness	Min. Height of Guard from Floor or Platform Level (ft, in.)
Woven Wire	Under 2 2-4 4-15	⅜ ½ 2	No. 16-⅜ in. No. 16-½ No. 12-2	8-0* 8-0 8-0
Expanded Metal	Under 4 4-15	½ 2	No. 18-½ in. No. 13-2	8-0 8-0
Perforated Metal	Under 4 4-15	½ 2	No. 20-½ in. No. 14-2	8-0 8-0
Sheet Metal	Under 4 4-15	 ...	No. 22 No. 22	8-0 8-0
Wood or Metal Strips Crossed	Under 4 4-15	⅜ 2	¾ in. wood or No. 16 metal	8-0
Wood or Metal Strips Not Crossed	Under 4 4-15	½ the width One width		
Plywood, Plastic or Equivalent	Under 4 4-15	...	¼ in. ¼ in.	8-0
Standard Railing	Min. 15 Max. 20	...	...	3-6

* Guards for rotating protruding objects should extend to a minimum height of 9 ft from the floor or platform.

From American National Standard B15.1, *Safety Code for Mechanical Power-Transmission Apparatus.*

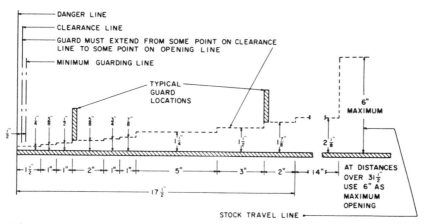

Point of operation guard locations. Barriers placed to touch dashed line will wedge hand or forearm. The danger line is the point of operation. The clearance line marks the distance required to prevent contact between the guard and the moving-parts. The minimum guarding line is ½ in. from the danger line.

From ANSI B11.1, *Safety Requirements for the Construction, Care, and Use of Mechanical Power Presses.*

machine, cause discomfort to the operator, or complicate the job of cleaning the area around the machine.

7. Be designed for the specific job and specific machine. Provisions must be made for oiling, inspection, adjusting, and repairing of the machine parts.
8. Be resistant to fire and corrosion. Be easy to repair.
9. Be strong enough to resist normal wear and shock, and durable enough to serve over a long period with minimum maintenance.
10. Not be a source of additional hazards, such as splinters, pinch points, sharp corners, rough edges, or other injury sources.

If possible, a safeguard covering rotating parts should be interlocked with the machine itself so that the machine cannot be operated unless the safeguard is in place. On some equipment this may be required by OSHA.

A (safe)guarding problem

When installations are contemplated, persons in the organization who have a specific interest in them should be consulted. The opinions of the operators of the machine, supervisors, setup and maintenance men,

TABLE XIV. POINT-OF-OPERATION PROTECTION

Type of Guarding Method	Action of Guard	Advantages	Limitations	Typical Machines on Which Used
		Enclosures or Barriers		
Complete, simple fixed enclosure	Barrier or enclosure which admits the stock but which will not admit hands into danger zone because of feed opening size, remote location, or unusual shape.	Provides complete enclosure if kept in place. Both hands free. Generally permits increased production. Easy to install. Ideal for blanking on power presses. Can be combined with automatic or semiautomatic feeds.	Limited to specific operations. May require special tools to remove jammed stock. May interfere with visibility.	Bread slicers Embossing presses Meat grinders Metal square shears Nip points of inrunning rubber, paper, textile, and other rolls Paper corner cutters Power presses
Warning enclosures (usually adjustable to stock being fed)	Barrier or enclosure admits the operator's hand but warns him before danger zone is reached.	Makes "hard to guard" machines safer. Generally does not interfere with production. Easy to install. Admits varying sizes of stock.	Hands may enter danger zone—enclosure not complete at all times. Danger of operator not using guard. Often requires frequent adjustment and careful maintenance.	Band saws Circular saws Cloth cutters Dough brakes Ice crushers Jointers Leather strippers Rock crushers Wood shapers
Barrier with electric contact or mechanical stop activating mechanical or electric brake	Barrier quickly stops machine or prevents application of injurious pressure when any part of operator's body contacts it or approaches danger zone.	Makes "hard to guard" machines safer. Does not interfere with production.	Requires careful adjustment and maintenance. Possibility of minor injury before guard operates. Operator can make guard inoperative.	Dough brakes Flat roll ironers Paper box corner stayers Paper box enders Power presses Rubber and paper calenders Rubber mills
Enclosure with electrical or mechanical interlock	Enclosure or barrier shuts off or disengages power and prevents starting of machine when guard is open; prevents opening of the guard while machine is under power or coasting. (Interlocks should not prevent manual operation or "inching" by remote control.)	Does not interfere with production. Hands are free; operation of guard is automatic. Provides complete and positive enclosure.	Requires careful adjustment and maintenance. Operator may be able to make guard inoperative. Does not protect in event of mechanical repeat.	Dough brakes and mixers Foundry tumblers Laundry extractors, driers, and tumblers Power presses Tanning drums Textile pickers, cards

ployees from all possible contact. Barriers should be provided for all similar danger points within 3½ ft of elevated walks or platforms in a horizontal direction, and within 9 ft of them in a vertical direction. The underside of belts running over passageways or work areas should be protected by screening so that they can do no harm if they break.

Enclosures should be made removable, or provided with hinged panels, and interlocked to facilitate inspection, oiling, or repairing of the parts of the mechanism, but should be taken off only when the machine is stopped, and the power disconnect is locked in an OFF position, and preferably only by the supervisor's special order. Machines should never be operated until the enclosures have been properly replaced and secured.

Ends of shafting projecting into passageways or into work areas should be cut off or protected by nonrotating caps or safety sleeves. Hangers should be securely fastened and well lubricated. Setscrews with slotted or hollow heads should be used instead of the projecting type. Couplings should be protected to prevent contact not only with coupling but the exposed rotating shafts also.

Exposed gears should be entirely enclosed by a substantial barrier or protected in some other equally effective way. Sheet metal is preferable. The inrunning side of the gearing should be safeguarded with special care. Removable barriers should be equipped with interlocks and the power disconnect locked in an OFF position for inspection and oiling. Wood barriers are permissible only where excessive exposure to water or chemicals might cause rapid deterioration of metal barriers.

It is essential to enclose gearing in power presses because the large forces involved may cause overloading, which in turn may cause fatigue cracks to start in the shaft or gear. The fact that gears have split in half and have fallen indicates the need of strong enclosures.

Clutches, cutoff couplings, and clutch pulleys with projecting parts 9 ft or less from the floor should be enclosed by stationary safe guards. Some clutches within the machine may be considered "guarded by location," but if any possibility of contact exists, a complete enclosure should be provided.

Cutting or shearing mechanisms

The hazards of cutting or shearing mechanisms lie at the points where work is being done, and where the movable parts of the machine

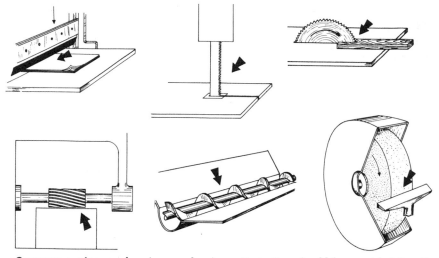

Common cutting or shearing mechanisms. Protection should be provided for all variations of such hazards.

approach or cross the fixed parts of the piece or machine. Guillotine cutters, shear presses, band and circular saws, milling machines, lathes, shapers, and abrasive wheels are typical machines which present cutting or shearing hazards.

A circular saw should be safeguarded by a hood which covers the saw at all times to at least the depth of the teeth. The hood should adjust itself automatically to the thickness of the material being cut and should remain in contact with it. The hood should be constructed so as to protect the operator from flying splinters or broken saw teeth.

A table saw should be equipped with a spreader or splinter. It should also have an anti-kickback device which will prevent material from being thrown back at the operator.

All portions of a band saw blade should be enclosed or otherwise safeguarded except the working side of the blade below the guide rules or gage. The enclosure for the portion of the blade between the sliding guide and the upper saw wheel guard should be self-adjusting if possible. Band saw and band knife wheels should be completely enclosed, and the barrier should be constructed of heavy material, preferably metal.

A swing cutoff saw should be equipped with a complete enclosure for the upper half of the saw. The lower half must be designed so it will ride over the fence and will drop down on the table or work being cut. Also a counterweight or other device should automatically return the saw to the back of the table when the saw is released at any point in its travel. Limit chains or other equally effective devices should be provided to prevent the saw from swinging past the table toward the operator.

Saw blades should be kept sharp at all times to prevent force feeding and the development of cracks.

Cutters on a milling machine should be shielded by an enclosure that provides positive protection for the operator's hands. A rotary cutter or slitter should have a barrier that completely encloses the cutting disks or knives to make it impossible for the operator to come into contact with the cutting edges while the machine is in motion.

The knife head on both hand- and power-operated shears should be equipped with a barrier which will keep the operator's fingers away from the cutting edge. The barrier should extend across the full width of the table and in front of the hold-down. · This barrier may be fixed or it may automatically adjust itself to the thickness of the material to be cut.

If material narrower than the width of the knife blade is being cut, adjustable finger barriers should be installed which will protect the open area at the sides of the material being sheared so that the operator cannot get his hands caught at the sides under the blade. The barrier may be slotted or perforated to allow the operator to watch the knife, but the openings should not exceed ¼-in. and should preferably be slotted vertically to provide maximum visibility. A cover should be provided over the entire length of the treadle on the shears, with only enough room between the cover and the treadle for the operator's foot. This device prevents accidental tripping of the machine if something should fall on the treadle.

Since the abrasive grinding wheel is a common power tool and since it is often used by untrained persons, it is the source of many injuries. Stands for grinding wheels should be heavy and rigid enough to prevent vibration and should be securely mounted on a substantial foundation. The wheels should not be forced on the spindle nor should they be too loose.

Installation of wheels should be in the charge of a person trained in correct and safe procedures. Before a wheel is mounted, it should be carefully examined for cracks or other imperfections which might cause the wheel to disintegrate.

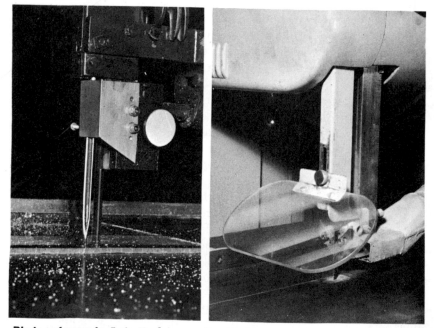

Blade safeguards. *Left:* Enclosure and guide for a 36-in. band saw. *Right:* Transparent plastic barrier on a band saw used for cutting metal.

The work rest should be rigid and should be set exactly ⅛-in. from the face of the wheel so that work cannot be caught between the wheel and the rest. Wheels should be kept true and in balance.

Each wheel should be enclosed with a substantial hood made of steel plate to protect the operator in case the wheel breaks. The threaded ends of the spindle should be covered so that clothing cannot get caught in them. Hoods should be connected to effective exhaust systems to remove chips and harmful dusts from the grinding operation.

The manufacturer's recommended wheel speed should be adhered to. Eye and face protection should be provided. Respirators should be used if considerable dust is generated, or if a poisonous or dangerous material is being worked on and the air contaminants cannot be properly exhausted.

Respirators should be frequently cleaned and sterilized, and one respirator should not be issued for use by several employees.

It is important that grinding wheels be properly stored and han-

Left: Interlocked finger barrier on metal shear must be down before shear can be operated. Black arrow (at bottom of picture) indicates locking cam.

Right: Grinding wheel must have (1) trued-up grinding face, (2) no visible cracks, and (3) the right-sized flange. Work rest is set ⅛ inch from face of wheel.

dled, to help prevent cracking or other damage to them. They should be stored in a dry place where they will not absorb moisture and should be kept in racks, preferably in vertical position. The manufacturers of the wheels can supply additional information on proper storage and handling methods.

Buffing and polishing wheels should be provided with exhaust hoods to catch particles thrown off by the wheels. Not only do hoods protect the operator—they also prevent accumulation of particles on the floor and in the area. Eye or face protection should be provided for operators. They should not wear gloves and loose clothing. Protruding nuts or the ends of spindles which might catch on the operator's hands or clothing should be covered by caps or sleeves.

Wire brush wheels. The same machine setup and conditions that apply to polishing and buffing wheels apply to brushes. The speed recommended by the manufacturer should be followed.

The hood on scratch wheels should enclose the wheel as completely as the nature of the work allows and should be adjustable so that the protection will not be lessened as the diameter of the wheel decreases. The hood should cover the spindle end, nut, and flange projection.

Personal protective equipment is especially important in the operation of scratch wheels because wires can break off.

The work should be held at the horizontal center of the brush. The wire tips of the brush should do the work. Forcing the work into the brush only results in (a) merely wiping or dragging the wires across the work, with no increase in cutting action, (b) increased wire breakage, and (c) a tendency for the work to become snagged.

Small pieces should be held in a jig or fixture.

More details about abrasive wheels, buffers, and scratch brushes are given in the next chapter.

Inrunning nip points

Whenever two or more parallel shafts, whether in close contact or some distance apart, rotate in opposite directions, an inrunning nip point is formed. Objects or parts of the body may be drawn into this nip point and be crushed or mangled. Typical examples of nip points are found on rolling mills, calenders, chains and sprockets, conveyors, belts and pulleys, racks and pinions, and at points of contact between any moving body and a corresponding stationary body.

Nip points should be made inaccessible by fixed barriers or should be protected by instantaneous body-contact cutoff switches with automatic braking devices. The inrunning side of rolls like those used for corrugating, crimping, embossing, printing, or metal graining should be protected by a barrier so arranged that the operator can feed material to the machine without catching his fingers between the rolls or between the barrier and the rolls.

Calenders and similar rolls should be protected by a device so arranged that the operator can immediately stop the rolls at the feed point by means of a level, rod, or treadle. Otherwise, the nip should be guarded by an automatic electronic device which will stop the rolls when anything but stock approaches the intake points.

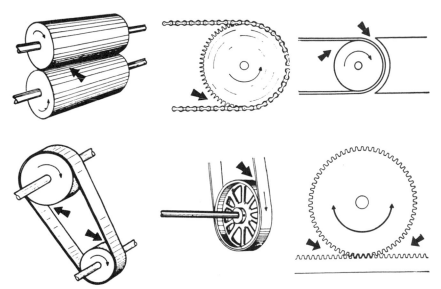

Typical inrunning nip points that require guarding.

Bar guards nip point of paper rewinder. Guard has adjustable cams so it always protects the inrunning nip point as diameter of the upper roll changes.

Enclosures are usually the most satisfactory way to protect chains and sprockets, racks and pinions, belts and pulleys (or sheaves), and drive mechanisms for conveyors.

Screw or worm mechanisms

The hazards involved in the operation of screw or worm mechanisms are the shearing action set up between the moving screw and the fixed parts of the machine and the mangling or battering action which will result if a person or part is caught in the mechanism. Screw or worm mechanisms are usually used for conveying, mixing, or grinding materials. Examples are: food mixers, meat grinders, screw conveyors, and dry material mixers and grinders of various types.

Covers should be provided for all mixers. The cover should be hinged to prevent removal and should have an interlock which will cut off the power source to stop action immediately when the cover is raised. Power switches for such machines should be locked out during maintenance or cleaning operations.

When screw conveyors are constantly fed while in motion and the use of an interlocked cover is impractical, a heavy screen or mesh guard or grid should be provided so that no opening is left when the cover is thrown back.

Rigid grids should cover the openings on grinder hoppers. Such coverings should be large enough to permit materials to be fed to the grinder, but small enough to prevent the operator's fingers from touching the cutting knives or the worm. Removable hoppers should be interlocked so that the grinders cannot be operated when the hoppers are removed.

Other safeguards for screw or worm mechanisms may include mechanical and electrical devices which require the operator to use both hands on the controls. Operating controls should be located so that they cannot be activated while any part of the body is in a position to be caught in the machine. The points of operation in many instances can be guarded by regulating the size, shape, and location of the feed opening.

Forming or bending mechanisms

The use of power presses, foot presses, and hand presses for stamping

308

and forming pieces of metal and other materials has grown rapidly. Hand and finger injuries on these presses, as commonly operated, have become so frequent that misuse or abuse of these machines constitutes one of the most serious sources of mechanical problems in accident prevention.

Factors that make the problem difficult are variations in operations

Hand-fed straightening machine has knee-level safety bar (A) that actuates emergency stop button. Rubber ring core (B) keeps operator's hand from reaching the rolls. Ring core is on a flexible mount to avoid creating additional pinch point. Because machine can be fed from either end, the same protection is at both ends. If the machine permitted continuous operation (such as in wire drawing) barrier (B) could be interlocked to stop machine in case the feed stock broke.

and operating conditions—in the size, speed, and type of press; in the size, thickness, and kind of pieces to be worked; in construction of dies; in degree of accuracy required in the finished work; and in length of run. It is foolish to simply depend on the skill of the operator to protect himself.

The supervisor who wants to make certain that operating methods are safe will have the die setter not only set the dies for a new run and

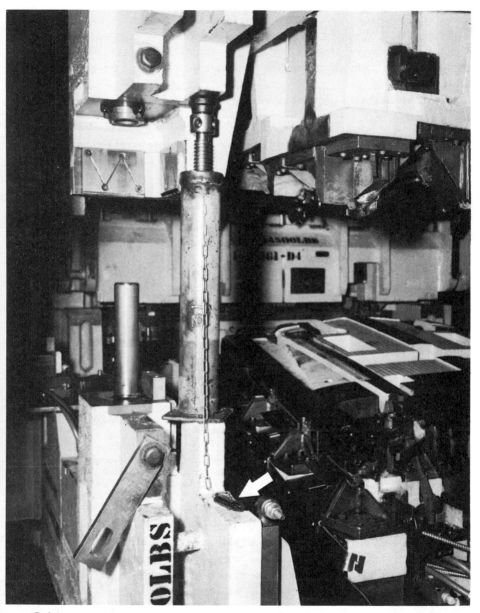

Safety prop for a power press protects operator when ram is raised. Before prop can be moved into position, electrical interlock plug (arrow) must be disconnected from the press; this deactivates the electrical circuit of the press.

test the machine for proper operation but also set and adjust the (safe)-guards. The supervisor then will have the operator run the machine a few strokes to make sure the (safe)guards are in place and the adjustments are correct. The supervisor should make sure the machine operator knows that each of these steps must be completed before the operator begins a production run.*

A procedure which omits any of these steps contains no guarantee that the dies are in alignment, that the kickout is working properly, that the clutch is in proper condition, and that the guard (if it is removable) is in place and operating. Foremen of long experience in hazard control customarily require the machine setup man to fill out a tag as written evidence that the procedure was thorough and complete for each setting.**

Each job poses questions. Whether fixed guards, pull-away devices, two-hand devices, or electric eye trips are to be used depends upon such factors as the kind of stock (piece, strip, roll), type of feed (hand, slide, dial, automatic), and type of knockout or ejection (mechanical or pneumatic).

Power presses sometimes repeat a cycle unexpectedly due to machine failure of one kind or another. The guard, therefore, must be designed to prevent injuries that could result from such occurrences. The most common action responsible for the loss of fingers is the impulsive reach for misplaced stock after the press has been tripped. Details of how to prevent these accidents are given later under "Automatic protection devices." Even the best training, experience, and supervision cannot substitute for well designed guards and the observance of safe practices.

---

* This procedure applies equally to shearing mechanisms; for example, guillotine cutters, veneer chippers, paper box corner cutters, leather dinking machines. As a group, these and similar machines present the worst kind of point-of-operation hazards, yet foremen often believe that (safe)guarding them would slow down production. The right (safe)guards, properly designed and utilized, will not only cover the hazard, but will usually increase production.

** OSHA Regulation §1910.217(e)(1) states, "Inspection, maintenance, and modification of presses—Inspection and maintenance records. It shall be the responsibility of the employer to establish and follow a program of periodic and regular inspections of his power presses to insure that all their parts, auxiliary equipment, and safeguards are in a safe operating condition and adjustment. *The employer shall maintain records of these inspections and the maintenance work performed.*"

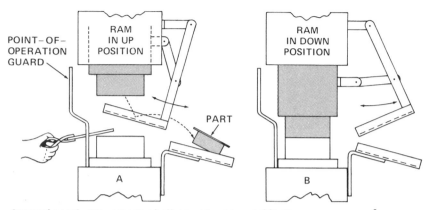

Approximately 80 percent of all power press accidents occur on secondary operations. Feeding can be made safe by adequate guarding of the point of operation, combined with gravity unload with mechanical-action chute. When ram goes up (A), formed part is knocked out and next blank is loaded. Mechanical-action chute moves out of the dies with downstroke of press (B).

Primary and secondary operations

Power press operations consist of primary operations and secondary operations.

In primary operations, the operator is not required to place his hands between the punch and the die. Examples are blanking, piercing, corner cutting, and other operations on long strip stock or in processing a large part. Primary operations should have die enclosures or fixed barrier guards.

In secondary operations, a preshaped part is further processed by being placed in a nest under the upper die. These operations include coining, drawing, and forming. Secondary operations, which account for some 80 percent of all power press accidents, should have the most effective protection within the limitations of the die—if possible, automatic or semiautomatic feeds and ejection and some form of protective guard.

Wherever possible, high-production or long-run dies should be guarded individually. This procedure saves the time involved in setup, and a permanent guard will not be detached or lost.

If die enclosures or fixed barriers cannot be used on secondary operations, an adjustable barrier device, a gate or movable barrier device, or a device which either prevents normal operation of the press until the hands are removed from the point of operation or the device itself removes the hands from the point of operation should be used. Unless the press has automatic or semiautomatic feeding and ejection with a die-enclosure or a fixed-barrier guard, hand tools should be used.

These devices must be maintained, and their use supervised strictly in accordance with the manufacturer's instructions. If they are not, they may fail. If they do fail, injuries may result and the workers' confidence in them will be destroyed.

Usually the insurance company safety engineer, the state factory inspector, or the local safety council engineer is willing to assist in the design and installation of the best guard for a specific operation.

A hand feeding tool is not a point-of-operation guard or protection device, and shall not be used instead of it.

Feeding methods

Automatic or semiautomatic feeds, if they can be used, are generally found to successfully (safe)guard the operation. In most cases, they also increase production or reduce costs. With these feeds, it is unnecessary for the operator to place his hand under the slide during ordinary feeding. He may, however, be tempted to do so if a piece sticks, or he may do so inadvertently. Therefore, it is necessary to provide a wood or soft metal stick or pick with which the operator can remove the material, if necessary, and an enclosure to prevent his putting a hand under the slide. The enclosure should be interlocked with the clutch brake mechanism.

The choice of feed will depend on the design of the die, the shape of the part being processed, the quantity of parts being processed, and the type of equipment available. Feeds include gravity or chute feeds, push feeds, follow feeds, magazine feeds, automatic magazine feeds, dial feeds, roll feeds, reciprocating feeds, hitch feeds, and transfer feeds. Feeds can be automatic, semiautomatic, or manual.

Among the advantages gained with automatic feeding are: (a) the operator does not have to reach into the point of operation to feed the press, (b) the feeding method usually makes it possible to enclose the die completely, and (c) the operator can load the feed mechanism, start the press, and then leave the vicinity of the press for a considerable number of strokes. Sometimes, the operator may be able to run several presses at once. Production volume is sufficient to justify the expense of automatic feeding.

With semiautomatic feeds, the operator does not have to reach into the point of operation, but usually must manually load the feed mechanism repeatedly or at frequent intervals. The feeding method usually makes it possible to enclose the die completely. Semiautomatic feeding is not adaptable for certain blanking operations or for nesting oddshaped pieces.

When manual feeding is required, some provision should be made that will eliminate by August 31, 1974, the need for the operator to place his hands or fingers within the point of operation. Special tools have been developed and used successfully on operations where automatic feeds or enclosure guards are impracticable. Such tools include pushers, pickers, pliers, tweezers, forks, magnets, and suction disks and are usually made of soft metal to protect the die. Strict discipline is necessary to enforce their constant use by the operator. Such hand tools are not to be considered as substitutes for guards, but should be used in conjunction with guard devices.

Bear in mind that few kinds of press guards provide complete protection. An automatic or semiautomatic feed may make it unnecessary for the operator to place his hands in the danger zone, but may not prevent his doing so. If the operator attempts to straighten a part just before it passes under the slide, this method of safeguarding provides no real protection. Therefore, good practice combines such guards with a two-hand trip (for actuating the clutch) that requires constant pressure or control during the downward stroke of the press.

Although automatic or semiautomatic methods of feeding can usually be installed on the majority of jobs that are fed manually, except certain blanking operations, automatic feeds should be supplemented by a substantial enclosure at the point of operation, especially on slow-moving equipment, for complete protection. If possible, machine parts should be adjusted to reduce the hazard. For instance, the stroke on a press may be limited so that the fingers cannot enter between the dies. More details are on pages 319 to 323.

Ejecting material

Safe removal of material is as necessary as safe placing of it. Since the way in which the finished piece is removed from the machine may influence the choice of feeding method, the removal method must be considered when an automatic or semiautomatic feed·is selected.

Various methods of ejection may be used: compressed air, punch, knockouts, strippers, and gravity. Operators should not be required to remove the finished parts from the die manually. Also, the operator should not be required to remove scrap from the die manually because of the hazards involved. Air blowoff systems, crankshaft-operated scrap cutters, and other devices may be used.

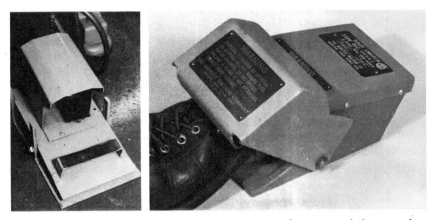

Foot pedals must be covered to prevent unintentional tripping of the press by the operator or by falling objects.

Foot controls

Power presses should have actuating devices which make sure that the operator's hands are out from under the slide when the press is operated. Such devices include two-hand switches or levers, treadle bars, pedals, and switches, located away from the point of operation. If two-hand switches or levers are used, relays or interlocks should be installed so that no switch or lever can be made inoperative and permit the press to be controlled with one hand. Pedals, foot switches, and pedal bars should be used only when it is absolutely necessary for the operator to have both hands occupied away from the point of operation when the press is operated.

Press brakes are the source of many accidents because the gaging stops are too low and the piece being processed slips beyond them.

Because the motion of the ram is slow, some operators reach through the area between the ram and the die to adjust the work and, in doing so, are caught. The proper type of starting device will make it impossible for the operator to reach under the ram after the press has begun to operate.

Foot controls should be covered by stirrup-type covers that extend over the entire length of the treadle arm (an inverted U-shaped metal shield above the control) to prevent accidental tripping. When two or more operators run a press, foot controls or hand controls should be connected in series so that each man must be in the clear before the press can be operated.

On die casting machines, two-hand tripping devices have been widely used, but many consider it safer to install a sliding door which covers the die area. As the door closes, protecting the hazardous zone, it activates switches that set the machine in motion. Such a door virtually eliminates burns from splashing material.

It is a prerequisite for a safe die casting operation that no operator be allowed to place his hands or arms between the dies at any time. Long-handled pliers or tongs or similar tools should be used to place and remove stock; mechanical feeds and ejectors are even safer.

Effective safeguards for these machines include two-hand tripping devices, sliding doors, treadle bars, and electrical or mechanical interlocking devices.

Part III—(Safe)guard Types and Maintenance

To eliminate the dangers involved in machine operation, either enclosures may be built and installed over the hazardous areas or the equipment may be redesigned to have no exposed dangerous parts.

The modern lathe is a good example of machinery made safe through improved design. Its motor drive and gear box are enclosed so that line shafts, pulleys, and belts are dispensed with. The modern power press, in which all the working parts with the exception of the slide (ram) are enclosed, is another good example.

(Safe)guards used to make machinery safe include the fixed guard or starters, the interlocking guard or barriers, and the automatic protection device. Automatic or semiautomatic feeding and ejection methods are also a means of safeguarding machine operations.

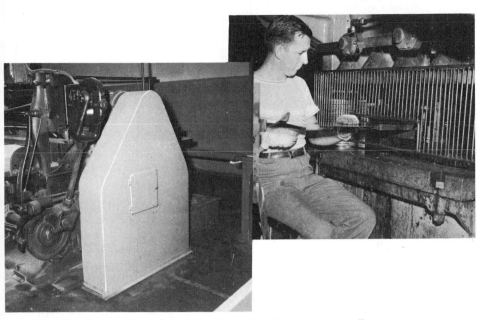

Fixed barriers. *Upper left:* Sheet metal encloses ends of rollers on a paper-cutter. Door in center of cover facilitates inspection and maintenance; the door must be interlocked. *Right:* Barrier guard for strip- and coil-fed stock has openings designed to keep fingers out of the danger area.

Fixed guards or enclosures

The fixed guard or enclosure is considered preferable to all other types and should be used in every case unless it has been definitely determined that this type is not at all practicable. The principal advantage of the fixed guard is that at all times it prevents access to the dangerous parts of the machine.*

Fixed safeguards may be adjustable to accommodate different sets of tools or various kinds of work. However, once they have been adjusted, they should remain "fixed"; under no circumstances should they be moved or detached.

Typical examples of the application of fixed (safe)guards are found on power presses, sheet leveling or flattening machines, milling machines, gear trains, drilling machines, and guillotine cutters.

* If a fixed barrier is to provide complete protection, the openings in it must be small enough to prevent a person from getting into the danger zone.

Some fixed barriers are installed at a distance from the danger point in association with remote feeding arrangements which make it unnecessary for the operator to approach the danger point.

Interlocking guards or barriers

Where a fixed (safe)guard cannot be used, an interlocking guard or barrier should be fitted onto the machine as the first alternative. Interlocking may be by mechanical, electrical, pneumatic, or a combination of types.

The purpose of the interlock is to prevent operations of the control that sets the machine in motion until the guard or barrier is moved into position so that the operator cannot reach the point of operation or the point of danger.

When the (safe)guard is open, permitting access to dangerous parts, the starting mechanism is locked to prevent accidental starting, and a locking pin or other safety device is used to prevent the basic mechanism from operating, for example, to prevent the main shaft from turning. When the machine is in motion, the enclosure cannot be opened. It can be opened only when the machine has come to rest or has reached a fixed position in its travel.

To be effective, an interlocking (safe)guard must satisfy three requirements:

1. It must (safe)guard the dangerous part before the machine can be operated.
2. It must stay closed until the dangerous part is at rest.
3. It must prevent operation of the machine if the interlocking device fails.

Two-hand tripping devices are incorporated in many types of interlocking controls. These devices require simultaneous and sustained pressure of both hands on switch buttons, air control valves, mechanical levers, or controls interlocked with foot control, to name just a few in order to start a cycle. Two-hand operating attachments should be so connected that it is impossible to block, tie down, or hold down one button, handle, or lever, and still operate the machine.

When gate devices or hinged barriers are used with interlocks, they should be so arranged that they completely enclose the pinch point or point of operation before the operating clutch can become engaged.

Access door on this expanded metal enclosure is interlocked—when door is opened for greasing or belt dressing, power is disconnected.

Interlocking controls are often installed on bakery machinery, guillotine cutters, power presses, dough mixers, some kinds of pressure vessels, centrifugal extractors, tumblers, and other machines on which covers or barricades must be in place before the starting control can be operated.

Automatic protection devices

An automatic protection device may be used, subject to certain restrictions, where neither a fixed barrier nor an interlocking (safe) guard is practicable. Such a device must prevent the operator from coming in contact with the dangerous part of the machine while it is in motion or must be able to stop the machine in case of danger.

An automatic device functions independently of the operator, and its action is repeated as long as the machine is in motion. The advantage of this type of device is that tripping can occur only after the operator's hands, arms, and body have been removed from the danger zone.

An automatic protection device is usually operated by the machine itself through a system of linkage, through levers, or by electronic means, and there are many variations. It may be a hand-restraint device or similar device, or a photoelectric relay.

Pull-away or hand-restraint devices are attached to the operator's hands or arms and connected to the slide, plunger, or outer side of the press in such a way that the operator's hands or fingers will be withdrawn from the danger zone as the slide, plunger, or outer slide descends. These devices should be readjusted at the start of every shift to make sure that they will properly pull the operator's hands clear of the danger zone.

All electronic safety devices for use on power presses are made to perform the same end function when energized—they act to interrupt the electric current to the power press (just as if the "stop" button had been pushed).

Electronic safety devices are effective only on power presses having air, hydraulic, or friction clutches. Such devices are not effective on power presses with positive clutches, because once the operating cycle of a power press starts, nothing can prevent completion of the cycle.

To be effective, the device should be operated from a closed electric circuit so that interruption of the current will automatically prevent the press from tripping. It is the supervisor's responsibility to make sure these devices are properly adjusted and maintained in perfect operating condition. Many injuries have occurred because of improper adjustments and because parts have been allowed to become worn or in need of repair.

One advantage claimed for this device is the absence of mechanism in front of the operator. It is particularly applicable on large presses. The electric-eye device should be installed far enough from the danger zone that it will stop the slide before the operator's hand

A fixed guard is not practical on this secondary operation. This automatic protection device will stop the ram if the operator's hand is in the danger zone when the press is activated. A tough transparent plastic allows an unobstructed view of the work. An interlock (not shown) prohibits press operation if, for some reason, the device is tampered with or is otherwise not operating properly.

can get under it, and enough light beams must be used to cover the open area with a curtain of light.

Indexing is another press shop term used to describe a mechanical method of feeding stock into press dies. One method is by the dial type of feeder. The dial feed is constructed, as its name implies, in the form of a dial having multiple stations that progress into the die by the indexing motion of the dial. The indexing of the dial should take place in conjunction with the up-stroke of the ram. When hand feeding, the index circuit should be controlled by dual-run buttons. Release of either button, during index, should stop the index cycle by releasing a safety clutch in the table. Safety guards should be connected to switches that stop the crank motion of the press whenever they are bumped. Both of these safety precautions will stop accidents should the operator be tempted to reach into the die area to correct an improperly positioned part on the dial.

Automation

Automation, a somewhat misused term, is defined here as the mechanization of processes by the use of automatically controlled conveying equipment.

Automation has minimized the hazards associated with manual moving of stock into and out of machines and transferring it from one machine to another. It has also minimized exposure to hernias, back injuries, and foot injuries.

Most finger and hand injuries result from the operator's exposing himself to the closing or working parts of a machine in the process of loading and unloading. The use of indexing fingers, sliding dies, and tongs or similar hand tools reduces the hazard from such exposure, but the supervisor still has the problem of seeing that these devices are used consistently and correctly.

Automation is a long step toward the solution of this problem since it completely eliminates the need for repetitive exposure at the point of operation. It utilizes automatic devices to move parts into and out of production equipment, turn them over, rotate them, shunt them to one side, remove scrap and waste, and perform other related functions.

Like most innovations, however, automation has brought not only benefits but also hazards. Since each automatic operation is dependent upon others, machine breakdown or failure must be corrected quickly. Because speed is highly important, a repairman may expose himself, inadvertently perhaps, to working parts of the equipment. It is therefore imperative to have a mandatory policy that equipment must be completely de-energized and locked out at the power source before servicing the machine.

Experience shows that those who work with automated equipment must have a thorough knowledge of the safety hazards and must be well trained in proper work methods and precautionary methods.

The simplest guarantee that equipment is safe to work on is to have the repairman or each member of a crew of repairmen place his own company-issued padlock on the power source. (Lockout procedures are discussed later in this chapter.)

Automation also has increased the use of stiles or crossovers. These should be constructed and installed in accordance with the standard.*

Automation eliminates or greatly reduces exposure to mechanical and handling hazards. In the single operation process, however, the basic principles of (safe)guarding of equipment must still be applied. These principles are:

1. Engineer the hazard out of the job insofar as possible,
2. (Safe)guard the remaining hazards,
3. Educate personnel, and
4. Insist on the use of (safe)guards provided.

Safe practices

(Safe)guards are of primary importance in eliminating machine accidents, but they alone are not enough. The employee who works around mechanical equipment or operates a piece of machinery must maintain a healthy respect for safeguards.

Before being permitted to run a piece of mechanical equipment, an operator should be instructed in all the safe practices required for safe operation of the machine. Even experienced operators should be given such training unless the supervisor is certain that they know the hazards and the necessary precautions to be taken. In addition, employees who do not themselves operate machinery, but who work in machine areas also should receive instruction in basic safe practices.

Positive procedures should be established to prevent misunderstandings, and the supervisor should enforce the following safe work practices:

1. No guard, barrier, or enclosure should be adjusted or removed for any reason by anyone unless that person has specific permission

---

* American National Standard A12.1, *Safety Code for Floor and Wall Openings, Railings, and Toe Boards.*

from the supervisor, he has been specifically trained to do this work, and machine adjustment is considered a normal part of his job.

2. Before safeguards or other guarding devices are removed so that repairs or adjustments can be made or equipment can be oiled or otherwise serviced, the power for the equipment must be turned off and the main switch locked out and tagged.

3. No machine should be started unless the (safe)guards are in place and in good condition.

4. Defective or missing (safe)guards should be reported to the supervisor immediately.

5. Employees should not work on or around mechanical equipment while wearing neckties, loose clothing, watches, rings, or other jewelry.

Maintenance of (safe)guards

The supervisor should see that inspection of machine (safe)guards is part of the regular schedule of machine inspection and maintenance. Such inspections are necessary because employees are inclined to operate their machines without (safe)guards if they are not functioning properly, if they have been removed for repairs, or if they interfere in any manner with their operations. A guard or enclosure that is difficult to remove or to replace may never be put back on again once it has been taken off. An inspection checklist can be developed for each type of machine to simplify the job and to provide a convenient record for followup.

(Safe)guarding for maintenance and repairs

Machines are subject to wear and deterioration through use and, as a result, may become unsafe to operate. Wear cannot be prevented, but it can be reduced to a minimum by control of the machine loads through proper manufacturing methods, alert supervision, attention of employees, and by good maintenance.

Lubrication is a basic maintenance function. Centralized lubrication will reduce the hazards to which the oiler is subjected in climbing ladders and in attempting to reach fairly inaccessible points. Changes can be made so that most lubrication can be done at floor level.

Drop enclosures, placed under conveyors, are made of low-cost, 11-gage galvanized fencing. Enclosures are easy to make, maintain, and interchange.

When oilers must get to the tops of presses to lubricate flywheel bearings, motors, and other parts, and since repairmen must likewise get to such places, good practice suggests that permanent ladders with sturdy enclosures (cages or wells) be installed.

Regardless of the type of lubrication necessary or the method used, it is the foreman's responsibility to know that his machinery—driving mechanisms as well as gears, motors, shafting hangers, and other parts—is being lubricated properly in its entirety. Where automatic lubrication is not possible or feasible, extension grease or oil pipes should be attached to the machine, so that the oiler can stay away from moving parts. However, this method should not be used where the oil or grease can congeal in the pipes.

Lockout procedure

Oilers and maintenance men often enter places "where no one ever goes" in order to do a job. As mentioned earlier in this chapter, even though apparatus may seem safe by virtue of its location, safeguards should be provided on moving parts in these out-of-way places. Another problem results when oilers or repairmen work on machines while they are operating, and fail to replace safeguards as soon as the work is completed. It is a definite responsibility of the supervisor to use all his authority to prevent such lapses in replacing guards, covers, or barriers.

Maintenance men, oilers, and repairmen should be provided with and use padlocks to lock out power-driven apparatus on which they

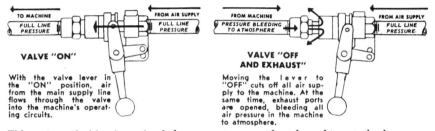

This automatic bleeder valve helps prevent an accidental machine stroke by permitting residual pressure to bleed immediately. Valve can be locked in either the "on" or the "off" position.

are to work, especially when the working place is some distance from the controls or is hidden from view. Each man should have his own lock and key, and no two locks or keys should be alike. Only locks bought by the company should be used. Locks can be painted different colors, according to craft, shift, or department, to facilitate identification. The employee's name or clock number should be stamped on his lock, or a metal tag bearing the owner's name should be attached to the lock.

Where a lockout system is to be set up, equipment must have built-in locking devices. They must be designed for the insertion of padlocks or have attachments on which locks can be placed. Methods include use of special tongues which hold several locks or sliding rods which can be extended and then locked to prevent operation of control handles. The lockout procedure, however, will be effective only if the supervisor sees that employees are trained to follow it, and then watches constantly for deviations. A typical lockout routine for maintenance and repair workmen to follow is given here:

1. Notify the operator that repair work is to be done on the machine.
2. Make sure the machine cannot be set in motion without your permission.
3. Place your own padlock on the power disconnect, even though another man's lock is already on it and blocks the mechanism. His lock will not protect you.
4. Place a MAN AT WORK sign at the control and block the mechanism. Make sure that neither the sign nor the blocking can be easily removed.
5. When the job is finished or your shift has ended, remove your own padlock and blocking. Never have another man remove it for you. Make sure first that you will not expose others to danger by removing your block or sign.
6. If you lose the key to your lock, report at once to the supervisor and get a new lock and key.

If two or more men are to work on a machine or piece of equipment where they might be separated, they should agree upon a foolproof set of signals. It is good practice to have a rule that when a machine or group of machines is stopped, only a specific, authorized person may start them again. In any case, the safe procedure is to give a warning and make sure that everyone is in the clear before power is turned on again.

It is good practice to have the supervisor check machines back into operation after repairs have been made and before new operations are undertaken. Every worker should be instructed not to start his machine unless (safe)guards are in place, and it is the supervisor's responsibility, not only to teach employees this practice, but to check to see that it is complied with constantly.

Replacements

Alerting management to the time when replacement of machinery and/ or its (safe)guards becomes advisable is partly, sometimes entirely, up to the foreman. Although it may seem that this is a production problem not closely related to accident prevention, such is not the case. When machinery, (safe)guards, and devices become worn to an extent that repairs cannot restore them to original operating efficiency, they constitute hazards, often of a concealed kind. For example, if a nonrepeat device on a press becomes so worn that it does not operate at all or fails to operate properly, amputation can result. These hazards can be eliminated only by replacement.

Several methods are used to determine when maintenance or overhaul is to be done to prevent breakdown, or when replacements are in order. Some supervisors rely on reports of the machine operators; some have occasional inspections made by oilers, machine setters, or similar mechanics; others put inspection on a time basis and do it themselves, or have it done, periodically. The best plan is a system of frequent, regularly spaced inspections.

(Safe)guarding hazards before accidents

Guarding the hazard is a fundamental principle of accident prevention not limited to machinery. The supervisor who undertakes to survey his department solely from the point of view of (safe)guarding the hazard—no matter what kind—can likely list a good many potential sources of injury which should be protected by barricades, rails, toe boards, enclosures, or other means.

As he lists electric switches, motors, engines, ladders, stairs, platforms, and pits, the supervisor might well ask himself, "Can an accident occur here, or here . . . or here?"

See Chapters 1 and 2 for details of eliminating accidents before they happen.

CHAPTER 11

Hand and Portable Power Tools

A major responsibility of supervisors is to train people to use hand and power tools safely. Employers are "responsible for the safe condition of tools and equipment used by employees, including tools and equipment which may be furnished by employees."—OSHA regulations § 1910.242(a).

This chapter discusses the problem of hand tool and portable power tool injuries. But to pin-point his own situation each supervisor should review his department or company accident records to picture his own problem, to find specific hazards, and to identify those employees who have an unusual number of injuries (and could use more training and closer supervision).

To be effective, a program to reduce tool injuries must include training in four basic safe practices.

I. Select the right tool for the job.

Do not permit unsafe practices, such as striking hardened faces of hand tools together (such as using a carpenter's hammer to strike another hammer, hatchet, or metal chisel), or using a file for a pry, a wrench for a hammer, and pliers instead of the proper wrench.

II. Keep tools in good condition.

Tools that have deteriorated to these conditions should not be used until repaired to meet factory specifications—wrenches with cracked or worn jaws, screw drivers with broken bits (points) or broken handles, hammers with loose heads, dull saws, and deteriorated extension cords or power cords on electric tools or broken plugs, and improper or removed grounding systems.

III. Use tools in the right way.

Screw drivers applied to objects held in the hand; knives pulled toward the body, and failure to ground electrical equipment are common causes of accidents.

IV. Keep tools in a safe place.

Many accidents have been caused by tools falling from overhead and by knives, chisels, and other sharp tools carried in pockets or left in tool boxes with cutting edges exposed.

Tool boxes containing personal tools can be neatly and safely stored in a storage rack like the one shown here. The shelves are movable so that employees can reach all sections safely to get their tool boxes out.

Control of Tool Accidents

A supervisor's program to control tool accidents should include the following activities:

1. Train employees to select the right tools for each job. Good job

safety analysis and job instruction training helps here. See Chapter 5, Instructing for Safety.

2. Establish regular tool inspection procedures (including inspection of employee-owned tools), and provide good repair facilities to ensure that tools will be maintained in safe condition.

3. Train and supervise employees in the correct use of tools for each job.

4. Establish a procedure for control of company tools, such as a check-out system at tool cribs.

5. Provide proper storage facilities in the tool room and on the job.

Each supervisor should also make a complete check of his operations to determine the need for special tools that will do the work more safely than ordinary tools. Special tools may require special handling and storage. For example, some tools (like powder-operated hand-tools) should be kept under lock and key.

Centralized tool control

Centralized tool control helps to assure uniform inspection and maintenance of tools by a trained man. The tool room attendant can help promote tool safety by recommending or issuing the right type of tool, by encouraging employees to turn in defective or worn tools, and by encouraging the safe use of tools.

The correct special personal protective equipment, such as welder's safety goggles or respirators, can be recommended and issued when the tool is distributed.

Centralized control facilities and the keeping of effective records on tool failure and other accident causes will help locate hazardous conditions and unsafe practices. Central storage facilities will also assure more positive control than will scattered storage.

A procedure should be set up so that the tool supply room attendant can send tools in need of repair to a department or manufacturer for a thorough reconditioning.

Some companies issue each employee a set of numbered checks which are exchanged for tools from the tool room. With this system the attendant knows where each tool is and can recall it for inspection at regular intervals.

Companies performing work at scattered locations may find that it is not always practicable to maintain a tool supply room. In such cases, the job supervisor should inspect all tools frequently and remove from service those found to be defective. Many companies have each supervisor check all tools weekly. A checklist can help systematize inspection.

Personal tools

In trades or operations where employees are required or prefer to have their own personal tools, the supervisor may encounter a serious problem of control. Personal tools are usually well maintained, but some persons may purchase cheap tools, make inadequate repairs, or attempt to use unsafe tools (such as hammers with broken and taped handles, or electrician's pliers with cracked insulation).

To remedy this, company policy should state the general requirements that tools must meet before they can be used; for example, they should meet the applicable American National Standard(s) and be listed by a nationally recognized testing firm. Then the supervisor should arrange a thorough inventory and initial inspection of personal tools, and should make sure he inspects and lists additional tools that employees buy just as though they were company-purchased tools. The supervisor must not permit the use of inferior grade tools.

Carrying tools

The workman should never carry tools which in any way might interfere with his using both hands freely on a ladder or while he is climbing on a structure. A strong bag, bucket, or similar container should be used to hoist tools from the ground to the job. Tools should be returned in the same manner, not brought down by hand, carried in pockets, or dropped to the ground.

Mislaid and loose tools cause a substantial portion of hand tool injuries. Tools are laid down on scaffolds, on overhead piping, on top of step ladders, and in other locations from which they can fall on persons below. Leaving tools overhead is especially hazardous where there is vibration or where men are moving or walking.

Chisels, screw drivers, and pointed tools should never be carried in a workman's pocket. They should be carried in a tool box or cart, in a carrying belt (sharp or pointed end down) like that used by electricians and steelworkers, in a pocket tool pouch, or in the hand

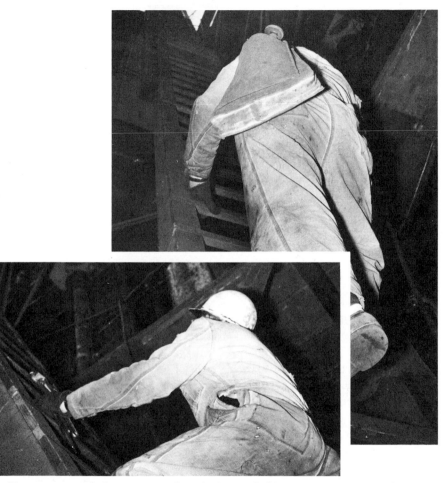

Never carry tools in the hand when climbing a ladder. Use a strong bag (right photo), or a bucket or similar container to hoist or lower tools

with points and cutting edges pointing away from the body.

Tools should be handed from one workman to another, never thrown. Edged or pointed tools should be passed, preferably in their carrying case, with the handle toward the receiver.

Workmen carrying tools on their shoulders should pay close attention to clearances when turning around and should handle the tools so that they will not strike other workmen.

Maintenance and Repair

The tool room attendant or tool inspector should be qualified by training and experience to pass judgment on the condition of tools for further use. Dull or damaged tools should not be returned to stock. Enough tools of each kind should be on hand so that when a defective or worn tool is removed from service, it can be replaced immediately with a safe tool.

Efficient tool control requires periodic inspections of all tool operations. These inspections should cover housekeeping in the tool supply room, tool maintenance, service, number of tools in the inventory, handling routine, and conditions of tools.

Responsibility for such periodic inspections is usually placed with the department supervisor and should not be delegated by him to the employee.

When metal tools break in use, there are usually detectable causes: overheating or underheating of the forging or steel when it was being hardened and tempered, cracks from improper forging, improper quenching, incorrect angle of cutting edge, or steel of poor quality.

Defects such as these will usually be found in tools of inferior construction, which, because of breakage and inefficiency, are more expensive in the long run than are tools of good quality. Therefore, it pays to purchase tools of the best quality obtainable.

Hand tools receiving the heaviest wear, such as chisels, wrenches, sledges, star-drills, blacksmith's tools, and cold cutters, require frequent maintenance on a regular schedule.

Proper maintenance and repair of tools require adequate facilities; workbenches, vises, a forge or furnace for hardening and tempering, safety goggles, repair tools, grinders, and good lighting. Workmen especially trained in the care of tools should be in charge of these facilities, otherwise tools should be sent out for repairs.

Tempering tools

Tools, such as chisels, stamps, punches, cutters, hammers, sledges, and rock drills, should be made of carefully selected steel and heat-treated so that they are hard enough to withstand blows without excessive mushrooming, and yet not be so hard that they chip or crack.

Combined effects of low temperatures and use finally broke this grounding clamp.

Hardening and tempering of tools requires special skills; therefore assign only experienced workmen. It is possible to temper a tool so that the surface is hard enough for cutting, yet the body is soft enough not to fracture.

Metal fatigue is a common problem in or near the polar regions. Metals exposed to sub-zero temperatures undergo "cold-soaking." A molecular change takes place that makes steel brittle and easily broken. Pneumatic impact tools, drill bits, and dies for threading conduit have their lives shortened.

Ferrous metals, including carbon steel, alloy steel, and cast iron, are characterized by decreased toughness and corresponding brittleness at low temperatures. Exceptions are the nine percent nickel-alloy steel and the chromium-nickel stainless steels.

Non-ferrous metals, such as aluminum and aluminum alloys

335

(nickel, copper and copper alloys, chromium, zinc and zinc alloys, magnesium alloys, and lead) are more resistant to low temperatures and are widely used in areas of extreme cold.

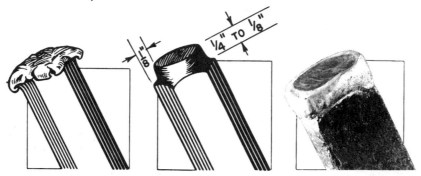

The hazard of burred heads can be reduced by safe-ending tools. This can be done by grinding or flame-cutting a shoulder on the tool and then bronze-welding it with a ⅛-inch rod.

Safe-ending tools

Tools, such as chisels, rock drills, flatters, wedges, punches, cold cutters, and stamping dies, should have heads properly hardened by a qualified workman. The hazard of burred heads can be reduced by use of a quick and economical procedure for safe-ending tools. A band of bronze is welded into a recess about ⅛ in. deep by ⅛ to ¼ in. wide that has been ground into the head of the tool.

Short sections of tight-fitting rubber hose can be set flush with the striking ends of chisels, hand drills, mauls, and blacksmith's tools to keep chips from flying, since they usually imbed themselves in the rubber sleeve.

Chisels, drift punches, cutters, and marking tools are available that are claimed not to spall or mushroom. This feature is attributed to a combination of alloys, plus scientific heat treatment.

Dressing tools

Tools require regular maintenance of their cutting edges or striking surfaces. In most cases, once the cutting or striking surfaces have been properly hardened and tempered, only an emery wheel, grindstone, or oilstone need be used to keep the tool in good condition.

Grind in easy stages. Keep the tool as cool as possible with water or other cooling medium.

Tools that require a soft or medium-soft head should be dressed as soon as they begin to mushroom. A slight radius ground on the edge of the head when it is dressed will enable the tool to stand up better under pounding and will reduce the danger of chips being knocked off.

A file or oilstone, rather than an abrasive wheel, is recommended for sharpening pike poles and axes.

A wood-cutting tool, because of its fine-cutting edge, should be sharpened on a grinder having a wheel recommended by the manufacturer for that type of tool. An oilstone set securely in a wood block

The correct way to whet a wood chisel or plane iron on a stone to produce a sharp cutting edge. The bevel should be placed on the stone with the back edge slightly raised.

placed on a bench should be used to obtain a fine, sharp cutting edge. The oilstone should never be held in the hand because a slip off the face of the stone could cause a severe hand injury. Often a few finishing strokes on a leather strop will produce a keener edge.

Metal cutting tools, because they generally have greater body, can be dressed or sharpened (or both) on an abrasive wheel. Care should be taken that the tool does not overheat from too much pressure against the wheel. The manufacturer's recommendations for type and kind of abrasive wheel should be followed. Each cutting edge should have the correct angle according to its use.

Handles

The handles of hand tools should be of the best straight-grained material, preferably hickory, ash, or maple, and should be free from

Tools with cracked or split handles are hazardous. Such handles should be replaced. Makeshift repairs, like winding wire or tape around a defective handle, should not be permitted. Through training and regular inspections, the supervisor should make sure men do not use tools in this condition.

slivers. Make sure that they are properly attached, because poorly fitted or loose handles are unsafe, may damage the material, and also make it difficult for the worker to control the tool.

No matter how tightly a handle may be wedged at the factory, both use and shrinkage will loosen it. Tool inspection should disclose damaged and loose handles. These should be removed from service and repaired, if feasible.

Use of Hand Tools

The misuse of common hand tools is a prolific source of injury to the industrial worker. In many instances, injury results because it is assumed that "anybody knows how" to use common hand tools. Observation and the record of injury show that this is not the case.

Supervisors should study each job and train new and retrain old employees on correct procedures for using tools. Where employees have the privilege of selecting or providing tools, the supervisor should advise them on the hazard of the job and insist on the safest equipment for each job.

Supervisors should enforce all rules. They should frequently check the condition of tools to be sure that they are maintained and sharpened correctly and that guards are not altered or removed. The use and the condition of personal protective equipment should be checked frequently.

Specific rules should be adapted for using hand tools in each operation. So important is this training that considerable attention is given, in the following pages, to these safe practices.

Metal-cutting tools

Chisels. Factors determining the selection of a cold chisel are the materials to be cut, the size and shape of the tool, and the depth of the cut to be made. The chisel should be made heavy enough so that it will not buckle or spring when struck. For best results and for safety, flat and cape chisels should be ground so that the faces form an angle of 70 degrees for cast iron, 60 degrees for steel, 50 degrees for brass, and about 40 degrees for babbitt and other soft metals.

A chisel only large enough for the job should be selected so that the blade is used rather than the point or corner. Also, a hammer heavy enough to do the job should be used.

Some workmen prefer to hold the chisel lightly in the hollow of the hand with the palm up, supporting the chisel by the thumb and first and second fingers. If the hammer glances from the chisel, it will strike the soft palm rather than the knuckles. Other workmen think that a grip with the fist holds the chisel more steady and minimizes the chances of glancing blows. Moreover, in some positions, this is the only grip that is natural or even possible. Hand protection can consist of a rubber pad, forced down over the chisel to provide a hand cushion.

When shearing with a cold chisel, the workman should hold the tool at the vertical angle that permits one bevel of the cutting edge to be flat against the shearing plane.

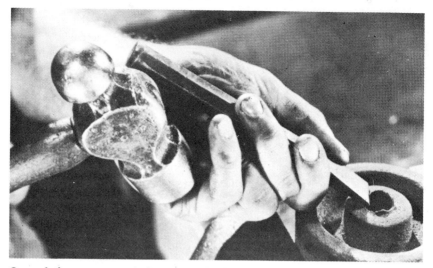

One of the recommended ways to hold a chisel—If hammer misses head of chisel, it will hit palm of workman's hand, not his knuckles.

Workmen should wear safety goggles when using a chisel and should set up a shield or screen to prevent injury to other workmen from flying chips. If a shield does not afford positive protection to all exposed employees, then glasses with side protection should be worn.

Bull chisels held by one man and struck by another require the use of tongs or a chisel holder to guide the chisel so that the workmen will not be exposed to injury. Both workmen should wear safety goggles and a safety hat. The man swinging the sledge should not wear gloves.

Tap and die work requires certain precautions. The work should be firmly mounted in the vise. Only a T-handle wrench or adjustable tap wrench should be used. Steady downward pressure should be used on the taper tap. Excessive pressure causes the tap to enter the hole at an angle or bind the tap, causing it to break. Too small a drill hole or lack of lubricant may also cause the tap to break. The hands should be kept away from broken tap ends. Broken taps should be removed with a tap extractor. If a broken tap is removed by using a prick punch or a chisel and hammer, the worker should wear safety goggles. When threads are being cut with a hand die, the hands and arms should be kept clear of the sharp threads coming through the die and metal cuttings should be cleared away with a brush.

TABLE XV SELECTOR FOR HACK SAW BLADES		
Pitch of Blade (Teeth per Inch)	*Stock To Be Cut*	*Explanation*
14	Machine Steel Cold Rolled Steel Structural Steel	The coarse pitch makes saw free and fast cutting
18	Aluminum Babbitt Tool Steel High Speed Steel Cast Iron	Recommended for general use
24	Tubing Tin Brass Copper Channel Iron Sheet Metal (18 gage or over)	Thin stock will tear and strip teeth on a blade of coarser pitch
32	Small Tubing Conduit Sheet Metal (less than 18 gage)	

Hack saws should be adjusted in the frame to prevent buckling and breaking, but should not be tight enough to break off the pins that support the blade. Install blade with teeth pointing forward.

The preferred blade to be used is shown in Table XV.

Pressure should be applied on the forward stroke, not upon the back stroke. If the blade is twisted or too much pressure is applied, the blade may break and cause injury to the hands or arms of the user.

Files. Selection of the right kind of file for the job will prevent injuries, lengthen the life of the file, and increase production. Inasmuch as the extremely hard and brittle steel of the file chips easily, the file should never be cleaned by being struck against a vise or other metal object—a file-cleaning card or brush should be used.

341

For the same reason a file should not be hammered or used as a pry. Such abuse frequently results in the file's chipping or breaking, causing injury to the user. A file should not be made into a center punch, chisel, or any other type of tool because the hardened steel may fracture in use.

The correct way to hold a file for light work is to grasp the handle firmly in one hand and use the thumb and forefinger of the other to guide the point. This technique will give good control and ensure better and safer work.

A file should never be used without a smooth, crack-free handle; otherwise, if the file should bind, the tang may puncture the palm of

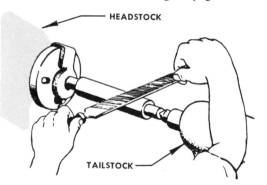

Although filing on a lathe is not recommended (an accurately ground tool bit and careful workmanship will keep the amount of filing to a minimum), the recommended procedure is to grip the file handle in the left hand and use the right hand to hold the top of the file. Be sure arm is kept out of range of face plate and lathe dog or chuck.

the hand, the wrist, or other part of the body. Under some conditions, a clamp-on, raised offset handle may be useful to give extra clearance for the hands. Files should not be used on lathe stock turning at high speed (faster than three turns per file stroke) because the end of the file may strike the chuck, dog, or face plate and throw the file (or metal chip) back at the operator hard enough to inflict serious injury. To help avoid contact with the turning parts of the lathe, the operator should always cross file.

Tin snips should be heavy enough to cut the material so easily that the workman needs only one hand on the snips and can use the other to hold the material. The material should be well supported before the last cut is made so that the cut edges do not press against the hands. When cutting long sheet metal pieces, push down the sharp ends next to the hand holding the snips.

Jaws of snips should be kept tight and well lubricated.

Workmen should wear safety goggles when trimming corners or slivers of sheet metal, because small particles often fly with considerable force. They should always wear gloves.

Cutters used on wire, reinforcing rods, or bolts should have ample capacity for the stock; otherwise, the jaws may be sprung or spread. Also, a chip may fly from the cutting edge and injure the user.

Cutters are designed to cut at right angles only. They should not be "rocked," hammered, or pushed against the floor to facilitate the cut because they are not designed to take the resulting strain. This practice will nick the cutting edges and result in future problems. A good rule is, "If it doesn't cut with ease, use a larger cutter or use a cutting torch."

Cutters require frequent lubrication. To keep cutting edges from becoming nicked or chipped, cutters should not be used as nail pullers or pry bars.

Cutter jaws should have the hardness specified by the manufacturer for the particular kind of material to be cut. By adjustment of the bumper stop behind the jaws, cutting edges should be set to have a clearance of 0.003 in. when closed.

Wood-cutting tools

Edged tools should be used so that, if a slip should occur, the direction of force will be away from the body. For efficient and safe work, edged tools should be kept sharp and ground to the proper angle. A dull tool does a poor job and may stick or bind. A sudden release may throw the user off balance or cause his hand to strike an obstruction.

Dressing of wood-cutting tools was discussed under that heading earlier in this chapter.

Wood chisels. Inexperienced employees should be instructed in the proper method of holding and using chisels. Handles should be free of splinters. The wood handle of a chisel struck by a mallet should be protected by a metal or leather cap to prevent its splitting.

The work to be cut should be free of nails to avoid damage to the

343

blade or cause a chip to fly into the user's face or eye.

The steel in a chisel is hard so that the cutting edge will hold, and is therefore brittle enough to break if the chisel is used as a pry.

When not in use, the chisel should be kept in a rack, on a work-

Metal, fiber, or heavy cardboard guards can be made to fit over sharp edges of tools. Tool boxes should be checked regularly to see that tools are stored with sharp edges protected.

bench, or in a slotted section of the tool box so that the sharp edges will be out of the way and not become damaged.

Saws should be carefully selected for the work they are to do. For fast crosscut work on green wood, a coarse saw (4 to 5 points per inch) should be used. A fine saw is better for smooth, accurate cutting in dry wood. Saws should be kept sharp and well set to prevent binding, and when not in use should be kept in racks.

Axes. To use an axe safely, workmen must be taught to check the axe head and handle, clear the area for an unobstructed swing, and swing correctly and accurately. Accuracy is attained by practice and proper hand hold—and good supervisory training. All others must keep a safe distance away.

A narrow axe with a thin blade should be used for hard wood, and a wide axe with a thick blade for soft wood. A sharp, well-honed axe yields better chopping speed and is much safer to use because it bites

Dull saws can be frustrating. In the rear is a saw that is in good condition. The one in front has been neglected—teeth are flattened, dull, and badly out of set.

into the wood. A dull axe will often glance off the wood being cut and strike the user in the foot or leg.

An axeman should make sure that he has a clear circle in which to swing his axe before he starts chopping. Also, he should remove all vines, brush, and shrubbery within the range, especially overhead vines that may catch or deflect his axe.

Axe blades should be protected with a sheath or metal guard wherever possible. When the blade cannot be guarded, it is safer to carry the axe at one's side. The blade on a single-edged axe should be pointed down.

Hatchets are used for many purposes and frequently cause injury. For example, when a workman attempts to split a small piece of wood while holding it in his hands, he may strike his fingers, hand, or wrist. Hatchets are dangerous tools in the hands of the inexperienced.

To start the cut, it is a good practice to strike the wood lightly with the hatchet, then force the blade through by striking the wood against a solid block of wood.

Hatchets should not be used for striking hard metal surfaces, since the tempered head may injure the user or others by flying chips. When using a hatchet for cutting or for driving nails in a crowded area, workmen should take special care to prevent injury to themselves and other workers.

Using a hatchet to drive nails is a poor practice. If used, however, the face of a hatchet used for driving nails should be square. Some companies prefer to use a corrugated face to prevent nails from flying.

Miscellaneous cutting tools

Planes, scrapers, bits, and drawknives should be used only by experienced men. These tools should be kept sharp and in good condition. When not in use, they should be placed in a rack on the bench, or in a tool box in such a way that will protect the user and prevent damage to the cutting edge.

Knives are more frequently the source of disabling injuries than any other hand tool. In the meatpacking industry, hand knives caused more than 15 percent of all disabling injuries, which is more than caused by any other agency.

The principal hazard in the use of knives is that the hands may slip from the handle onto the blade or that the knife may strike the body or the free hand. A handle guard or a finger ring (and swivel) on the handle eliminates these hazards. Adequate guarding is important.

The cutting stroke should be away from the body. If that is not possible, then the hands and body should be in the clear, a heavy leather apron or other protective clothing should be worn, and, where possible, a rack or holder should be used for the material to be cut. Jerky motions should be avoided to help maintain balance. Be sure employees are trained and supervised. See picture on page 177.

Belt repairmen and other workmen who must carry knives with them on the job should keep them in sheaths or holders. Never carry a sheathed knife on the front part of a belt—always carry it over the right or left hip, toward the back. This will prevent severing a leg artery or vein in case of a fall.

Knives should never be left lying on benches or in other places where they may cause hand injuries. When not in use, they should be kept in racks with the edges guarded. Safe placing and storing of

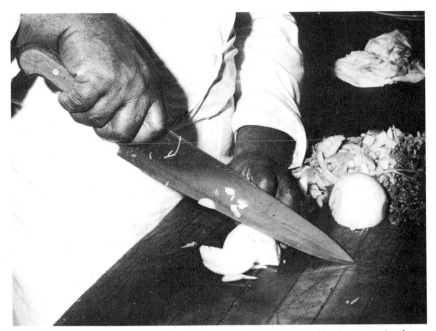

Slice vegetables with a rocking motion, rather than chopping or hacking away. Hold the point of the chef's knife against the cutting surface. Fingers that hold the object should be curled back and in the clear.

knives is important to knife safety.

Ring knives—small, hooked knives attached to a finger ring—are used where string or twine must be cut frequently. Supervisors should make sure that the cutting edge is kept outside the hand, not pointed inside. A wall-mounted cutter or blunt-nose scissors would be safer.

Carton cutters are safer than hooked or pocket knives for opening cartons. They not only protect the user, but eliminate deep cuts that could damage carton contents. Frequently, damage to contents of soft plastic bottles may not be detected immediately; subsequent leakage may cause chemical burns, damage other products, or start a fire.

To cut corrugated paper, a hooked linoleum knife permits good control of pressure on the cutting edge and eliminates the danger of the blade suddenly collapsing as in the case of a pocket knife. Be sure hooked knives are carried in a pouch or heavy leather (or plastic) holder. The sharp tip must not stick out.

Supervisors should make certain that employees who handle

knives have ample room in which to work so they are not in danger of being bumped by trucks, the product, overhead equipment, or other employees. For instance, a left-handed worker should not stand close to a right-handed person; the left-handed person might be placed at the end of the bench or otherwise given more room. Workers should be trained to cut away from or out of line with their bodies.

Supervisors should be particularly careful about the hazard of employees leaving knives hidden under product, under scrap paper or wiping rags, or among other tools in work boxes or drawers. Knives should be kept separate from other tools to protect the cutting edge of the knife as well as to protect the employee.

Work tables should be smooth and free of slivers. Floors and working platforms should have slip-resistant surfaces and should be kept unobstructed and clean. If sanitary requirements permit mats or wooden duck boards, they should be in good repair, so workers do not trip or stumble. Conditions which cause slippery floors should be controlled as much as possible by good housekeeping and frequent cleaning.

Careful job and accident analysis may suggest some slight change in the operating procedure which will make knives safer to use. For instance, on some jobs special jigs, racks, or holders may be provided so it is not necessary for the operator to stand too close to the piece being cut.

The practice of wiping a dirty or oily knife on the apron or clothing should be discouraged. The blade should be wiped with a towel or cloth with the sharp edge turned away from the wiping hand. Sharp knives should be washed separately from other utensils and in such a way that they will not be hidden under soapy wash water.

Horseplay should be prohibited around knife operations. Throwing, "fencing," trying to cut objects into smaller and smaller pieces, and similar practices are not only dangerous but reflect inadequate supervision.

Supervisors should make sure that nothing is cut that requires excessive pressure on the knife—such as, frozen meat. Food should be thawed before it is cut or else it should be sawed. Knives should not be used as a substitute for can openers, screwdrivers, or ice picks.

Ice picks. Serious hand injuries occur frequently when ice is held in the hand and breaks unexpectedly. To ensure safety in the use of an ice pick, first place the pieces of ice on a flat surface.

The handle of an ice pick should be flat or square, or have at least one flat surface to prevent the pick from rolling. Ice picks should be carried in sheaths or stored in a holder that covers the points, and should not be used as scribes or pin punches.

Torsion tools

Socket wrenches are safer to use than adjustable or open-end wrenches.

Open-end or box wrenches should be inspected to make sure that they fit properly and should never be used if jaws are sprung

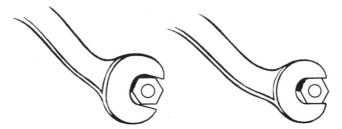

Wrench jaws that fit (and are not sprung or cracked) prevent damage to the head of the nut, and are not likely to slip and cause injury to the user.

or cracked; when defective they should be taken out of service until repaired. It is dangerous to use a piece of pipe or a wrench for added leverage, or to use a wedge to make a wrench fit.

Socket wrenches give great flexibility in hard-to-reach places. The use of special types should be encouraged where there is danger of injury.

Adjustable wrenches are used for many purposes. They are not intended, however, to take the place of standard open-end, box, or socket wrenches. They are used mainly for nuts and bolts that do not fit a standard wrench. Pressure is always applied to the fixed jaw.

Pipe wrenches. Workmen, especially those on overhead jobs, have been seriously injured when pipe wrenches slipped on pipes or fittings, causing the men to lose their balance and fall. Pipe wrenches, both straight and chain tong, should have sharp jaws and be kept clean to prevent their slipping.

The adjusting nut of the wrench should be inspected frequently. If it is cracked, the wrench should be taken out of service. A cracked nut may break under strain, causing complete failure of the wrench and possible injury to the user.

Using a wrench of the wrong length is also a source of accidents. A wrench handle too small for the job does not give proper grip or leverage. An oversized wrench handle may strip the threads or break the fitting or the pipe suddenly, causing a slip or fall.

A piece of pipe ("cheater") slipped over the handle to give added

The unsafe practice of using a hammer to force a wrench can result in serious injury if the wrench slips or breaks, and always damages the wrench by springing or spreading the jaws. Employees should be trained to select a wrench of the proper size for the particular application.

leverage also can strain a pipe wrench to the breaking point. The handle of every wrench is designed to be long enough for the maximum allowable safe pressure.

A pipe wrench should never be used on nuts or bolts, the corners of which will break the teeth of the wrench, making it unsafe to use on pipe and fittings. Also, a pipe wrench, when used on nuts and bolts, also damages their heads. A pipe wrench should not be used on valves, struck with a hammer, nor used as a hammer unless, as with specialized types, it is specifically designed for such use.

Tongs usually are bought, but some companies make their own to perform specific jobs. Often they are poorly designed—hands and fingers are pinched when the tongs are closed.

To prevent pinching, the end of one handle should be up-ended toward the other handle, to act as a stop. It is also possible to braze, weld, or bolt bumpers on the handles a short distance behind the pivot point so that the handles cannot close against the fingers.

Pliers. Side-cutting pliers sometimes cause injuries when short ends of wire are cut. A guard over the cutting edge and the use of safety glasses will help prevent eye injuries.

The handles of electricians' pliers should be insulated. In addition, the men should wear electricians' gloves if they are to work on energized lines.

Because pliers do not hold the work securely, they should not be used as a substitute for a wrench.

Special cutters include those for cutting banding wire and strap. Claw hammers and pry bars should not be used to snap metal banding material. Only cutters designed for the work provide safe and effective results.

Nail band crimpers make it possible to keep the top band on kegs and wood barrels after nails or staples have been removed. Use of these tools eliminates injury caused by reaching into kegs or barrels that have projecting nails and staples.

Pipe tongs should be placed on the pipe only after the pipe has been lined up and is ready to be made up. A 3- or 4-in. block of wood should be placed near the end of the travel of the tong handle and parallel to the pipe to prevent injury to the hands or feet in the event the tongs slip.

Workmen should neither stand nor jump on the tongs nor place extensions on the handles to obtain more leverage. They should use larger tongs, if necessary, to do the job.

Screw drivers. The screw driver is probably the most commonly used and abused tool. The practice of using screw drivers for punches, wedges, pinch bars, or pries should be discouraged. If used in such manner, they become unfit for the work they are intended to do. Furthermore, a broken handle, bent blade, dull or twisted tip may cause a screw driver to slip out of the slot and cause a hand injury.

Cross-slot (Phillips-head) screw drivers are safer than the square bit type, because they have less tendency to slip. The pressure on the tip is more evenly distributed and results in less wear on the tip. The tip must be kept clean and sharp, however, to permit a good grip on the head of the screw.

A screw driver tip should fit the screw. A sharp square-edged

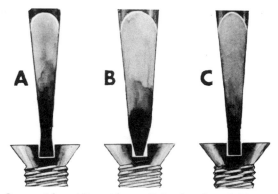

Screw driver bits with parallel sides shaped like in illustration **C,** or undercut slightly like in **A,** will grip the slot more securely than the bit shown in **B.**

bit will not slip as easily as a dull, rounded one, and requires less pressure.

The part to be worked upon should never be held in the hands; it should be laid on a bench or flat surface, or held in a vise. This practice will lessen the chance of injury to the hands if the screw driver should slip from the work.

No screw driver used for electrical work should have the blade or rivet extending through the handle. Also, both blade and handle should be insulated except at the tip.

Shock tools

A hammer should have a securely wedged handle suited to the type of head used. The handle should be smooth, free of oil, shaped to fit the hand, and of the specified size and length. Employees should be warned against using a steel hammer on hardened steel surfaces. Instead, a soft-head hammer or one with a plastic, wood, or rawhide head should be used. Safety goggles should be furnished and worn to protect against flying chips, nails, or scale.

The chipping or spalling of a hammer varies with four things:

1. The squarer the corners of the hammer the easier it chips.
2. The harder the hammer is swung the more likely it is to chip.
3. Chipping increases as the hardness of the object being struck increases.
4. The greater the angles between the surface of the object and the hammer face, the greater are the chances of chipping.

Selection of the proper hammer is important. One that is too light is as unsafe and inefficient as one that is too heavy.

Sledge hammers can have two common unsafe conditions: split handles and loose or chipped heads. Because these tools are used infrequently in some industries, the heads may become loose or chipped and the defect not be noticed.

Some companies place a steel band around the head and bolt it to the handle to prevent the head from flying off. The heads should be dressed whenever they start to check or mushroom.

A sledge hammer so light that it bounces off the work is hazardous; likewise, one too heavy is hard to control and may cause body strain.

Riveting hammers, often used by sheet metal workers, should have the same kind of use and care as ball peen hammers and should be watched closely for checked or chipped faces.

Carpenter's or claw hammers are designed primarily for driving and drawing nails. Their shape, depth of face, and balance make them unsuitable for striking objects, such as cold chisels.

The faces should be kept well dressed at all times to reduce the hazard of flying nails while they are being started into a piece of wood. A checker-faced head is sometimes used to reduce this hazard.

Eye protection is advisable for all nailers and all employees working in the same area, as in a shipping room, when blocking and bracing trucks or railway cars, or in carpenter shops and the like.

When a nail is to be drawn from a piece of wood, a block of wood may be used under the hammer head to increase the leverage.

Spark-resistant tools

So-called spark-resistant tools of nonferrous materials are sometimes advised for use where flammable gases, highly volatile liquids, and explosive materials are stored or used. The intensified sparks from steel tools are capable of igniting substances such as gunpowder, lint, TNT, carbon disulfide, and ethyl ether.

In certain circumstances, steel coated with aluminum paint can emit sparks when struck with a metal striker (steel, brass, or spark-resistant alloys), and such sparks may ignite mixtures of flammable gases or vapors in air. The sparks are due to the thermite reaction between iron oxide and aluminum and are, therefore, produced only in the presence of iron oxide. The extent of this hazard depends on the nature of the surface covered and the nature of the paint, and it is increased if the painted steel is subjected to heat.

Nonferrous tools reduce the hazard from sparking but do not eliminate it. They need inspection before each use to be certain that they have not picked up foreign particles which could produce friction sparks, thereby obviating the value of these special tools.

Portable Power Tools

Portable power tools are divided into four primary groups according to the power source: electric, pneumatic, gasoline, and explosive (powder actuated). Several types of tools, such as saws, drills, and grinders, are common to the first three groups; whereas explosive tools are used exclusively for penetration work and cutting.

A portable power tool presents similar hazards as a stationary machine of the same kind, in addition to the risks of handling. Typical injuries caused by portable power tools are burns, cuts, and strains. Sources of injury include electric shock, particles in the eyes, fires, falls, explosion of gases, and falling tools.

The power cord should always be disconnected before accessories on a portable tool are changed, and guards should be replaced or put in correct adjustment before the tool is used again.

A tool should not be left in an overhead place where there is a chance that the cord, if pulled, will cause the tool to fall. The cord and the tool may be suspended by counterweighted rope or string which keeps the cord out of the operator's way and also counterbalances some of the weight of the tool and the cord. Power extension cords on the floor create a stumbling or tripping hazard. They should be suspended over aisles or work areas, where possible. The cord should be suspended in such a way that it will not be struck by other objects or by material being handled or moved. An unexpected pull on the cord might cause the tool to jam or otherwise expose the operator to injury.

Do not hang power cords over nails, bolts, or sharp edges. Cords should be kept away from oil, hot surfaces, and chemicals.

Because of the extreme mobility of power driven tools, they can easily come in contact with the operator's body. At the same time, it is difficult to guard such equipment completely. There is the possibility of breakage because the tool may be dropped or roughly handled. Furthermore, the source of power (electrical, mechanical, air, or explosive cartridge) is brought close to the operator, thus creating additional hazards.

When using explosive cartridge equipment for driving anchors into concrete, or when using air-driven hammers or jacks, it is recommended that ear plugs, safety glasses, and/or face shields be used. All companies and manufacturers of portable power tools attach to each tool a set of operating rules or safe practices. These supplement the thorough training each powered tool operator should have received.

Power driven tools should be set in safe places and not left in areas where they may be struck by passers-by, and activated.

Selection

When you replace a hand tool with a power tool, you may be replacing a less serious hazard with a more serious one. Check with your safety professional to make sure the tools you select meet current safety standards. The tool manufacturer, too, can recommend the best tool to do the job you want. Don't just tell him the job you want accomplished, but also the material to be worked on, and the space available in the work area of your shop or plant. Tell if the operation is intermittent or continuous; too light a tool may fail or cause undue operator fatigue if used over an extended time.

Electric tools

Electric shock is the chief hazard from electrically powered tools. Types of injuries are electric flash burns, minor shock that may cause falls, and shock resulting in death.

Serious electric shock is not entirely dependent on the voltage of the power input. The ratio of the voltage to the resistance determines the current that will flow and the resultant degree of hazard. The current is regulated by the resistance to ground of the body of the operator and by the conditions under which he is working. It is possible for a tool to operate with a defect or short in the wiring, and the use of a ground wire protects the operator under all conditions.

Insulating platforms, rubber mats, and rubber gloves provide an additional factor of safety when tools are used in wet locations, such as in tanks or boilers and on wet floors.

Low voltage of 6, 12, 24, or 32 volts through portable transformers will reduce the shock hazard in wet locations.

Electric tools used in wet areas or in metal tanks expose the operator to conditions favorable to the flow of current through his body, particularly if he is wet with perspiration. Most electric shocks from tools have been caused by the failure of insulation between the current-carrying parts and the metal frames of the tools. Only tools in good repair and listed by Underwriters' Laboratories, Inc., should be used.

Double insulated tools. Protection from electric shock, while using portable power tools, has been described as depending upon third wire protective grounding. "Double insulated" tools, however, are available which provide more reliable shock protection without third-wire grounding. Paragraph 250–45 of the 1971 *National Electrical Code* permits "double insulation" for portable tools and appliances. Tools in this category are permanently marked by the words "double insulation" or "double insulated." Units designed to this category which have been tested and listed by Underwriters' Laboratories, Inc., will also employ the use of the UL symbol. Many U.S. manufacturers are also using the symbol

to denote "double insulation." This symbol is also being widely used in most European countries.

This double insulated or all-insulated tool does not require separate ground connections; the third wire or ground wire is not needed and should not be used.

Failure of insulation is harder to detect than worn or broken external wiring, and points up the need for frequent inspection and

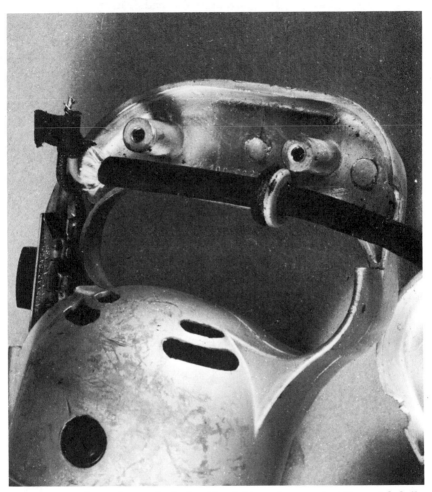

Exposed energized wire inside the handle of this tool can contact the metal shell. Tool will still run, but if the operator presents a good ground (such as by touching a water pipe) he would sustain electric shock. Many accidents primarily caused by electric shock are not reported as such and are charged against falls, tools dropped on feet, burns, and the like.

thorough maintenance. Care in handling the tool and frequent cleaning will help prevent the wear and tear that cause defects.

Grounding of portable electric tools provides the most convenient way of safeguarding the operator. If there is any defect or short circuit inside the tool, the current is drained from the metal frame through a ground wire and does not pass through the operator's body. All electric power tools should be effectively grounded except the double insulated and cordless types. Grounded tools are as safe as double insulated or low voltage tools.

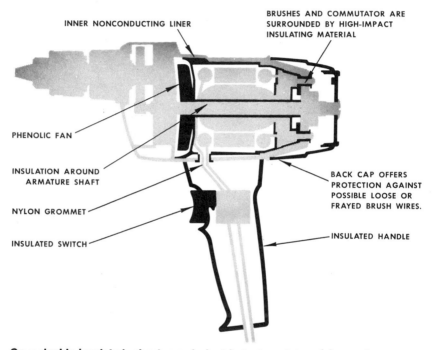

INNER NONCONDUCTING LINER

BRUSHES AND COMMUTATOR ARE SURROUNDED BY HIGH-IMPACT INSULATING MATERIAL

PHENOLIC FAN

INSULATION AROUND ARMATURE SHAFT

NYLON GROMMET

INSULATED SWITCH

BACK CAP OFFERS PROTECTION AGAINST POSSIBLE LOOSE OR FRAYED BRUSH WIRES.

INSULATED HANDLE

On a double-insulated, shock-proof electric tool, an internal layer of protective insulation isolates the electrical components from the outer metal housing.

The noncurrent-carrying metal parts of portable and/or cord- and plug-connected equipment required to be grounded may be grounded —either (*a*) By means of the metal enclosure of the conductors feeding such equipment, provided an approved grounding-type attachment plug is used, or (*b*) by means of a grounding conductor run with the power supply conductors in a cable assembly or flexible cord that is properly terminated in an approved grounding-type attachment plug having a fixed grounding contacting member. The grounding conductor may be uninsulated; if individually covered, however, it must be finished a continuous green color or a continuous green color with one or more yellow stripes.

(c) By special permission, nonportable cord- and plug-connected equipment can be grounded by a separate flexible wire or strap, insulated or bare, that has been protected (as well as practicable) against physical damage.

Electric cords should be inspected periodically and kept in good condition. Heavy-duty plugs that clamp to the cord should be used to prevent strain on the current-carrying parts if the cord is accidentally pulled. Employees should be trained not to jerk cords and to protect them from sharp objects, heat, and oil or solvents that might damage or soften the insulation.

Electric drills cause injuries in several ways: a part of the drill may be pushed into the hand, the leg, or other parts of the body, the drill may be dropped when the operator is not actually drilling, and the eyes may be hit either by material being drilled or by parts of a broken drill bit. Although no guards are available for drill bits, some protection is afforded if drill bits are carefully chosen for the work to be done, such as being no longer than necessary to do the work.

Where the operator must guide the drill with his hand, the drill should be equipped with a sleeve that fits over the drill bit. The sleeve protects the operator's hands and also serves as a limit stop if the drill should plunge through the material.

Oversized bits should not be ground down to fit small electric drills; instead, an adapter should be used that will fit the large bit and provide extra power through a speed reduction gear; however, this again is an indication of improper drill size. When drills are used, the pieces of work should be clamped or anchored to prevent whipping.

Electric saws are usually well guarded by the manufacturer, but employees must be trained to use the guard as intended. The guard should be checked frequently to be sure that it operates freely and encloses the teeth completely when it is not cutting, and the unused portion of the blade when it is cutting.

Circular saws should not be jammed or crowded into the work. The saw should be started and stopped outside the work. At the beginning and end of the stroke, or when the teeth are exposed, the portable power hand-held circular saw operator must use extra care to keep his body and power cord out of the line-of-cut. The safest saws have a trigger switch to shut off power when pressure is released.

Lower movable guard of this portable electric saw always returns to the guarded position. Operator should keep his fingers away from the trigger when saw is not being used.

Abrasive wheels, buffers, and scratch brushes should be guarded as completely as possible. For portable grinding, the maximum angular exposure of the periphery and sides should not exceed 180 degrees and the top half of the wheel should always be enclosed. Guards should be adjustable so that operators will be inclined to make the correct adjustment rather than remove the guard. However, the guard should be easily removable to facilitate replacement of the wheel. In addition to mechanical guarding the operator should wear safety goggles at all times in case the wheel disintegrates.

The portability of a grinding wheel exposes it to more abuse than that given a stationary grinder. The wheel should be kept away from water and oil, which might affect its balance; the wheel should be protected against blows from other tools; and care should be exercised not to strike the sides of a wheel against objects or to drop the wheel. Cabinets or racks will help protect the wheel against damage.

The speed and weight of a grinding wheel, particularly a larger one, make it more difficult to handle than some other power tools. Since part of the wheel must necessarily be exposed, it is important that employees be trained in the correct way to hold and use the wheel so that it does not touch the clothes or the worker's body.

The wheels should be mounted only by trained workmen, with the wheels and safety guards conforming to American National Standard B7.1, *Safety Code for the Use, Care, and Protection of Abrasive Wheels,* now promulgated in OSHA regulations as § 1910.243(e). Grinders should be marked to show the maximum abrasive wheel size and speed. Abrasive wheels should be sound-tested (ring-tested) before being mounted.

Sanders of the belt or disk type cause serious skin "burns" when the rapidly moving abrasive touches the body. Because it is impossible to guard sanders completely, employees require thorough training in their use. The motion of the sander should be away from the body, and all clothing should be kept clear of the moving parts. Dust-type safety goggles or plastic face shields should be worn and, if harmful dusts are created, a respirator approved by the National Institute for Occupational Safety and Health (NIOSH) for the exposure should be worn.

Sanders require especially careful cleaning because of the dusty nature of the work. If a sander is used steadily, it should be dismantled periodically, as well as thoroughly cleaned every day by being blown out with low-pressure air—less than 30 psig. If compressed air is used, the operator should wear safety goggles or work with a transparent chip guard between his body and the air blast.

Because wood dust presents a fire and explosion hazard, keep dust to a minimum; sanders can be equipped with a dust collection or vacuum bag. Electrical equipment should be designed to minimize the explosion hazard. Fire extinguishers approved for Class C (electrical) fires should be available. Employees should be trained in what to do in case of fire.

Soldering irons are the source of burns and of illness resulting from inhalation of fumes. Insulated, noncombustible holders will practically eliminate the fire hazard and the danger of burns from accidental contact. Ordinary metal covering on wood tables is not sufficient because the metal conducts heat and may ignite the wood.

Holders should be designed so that employees cannot accidentally touch the hot irons if they reach for them without looking. The best holder completely encloses the heated surface and is inclined so that the weight of the iron prevents it from falling out. Such holders must be well ventilated to allow the heat to dissipate, otherwise the life of the tip will be reduced and the wiring or printed circuit may be damaged. Also see NSC Data Sheet 445, *Hand Soldering and Brazing*.

Harmful quantities of fumes from lead soldering should not be allowed to accumulate. Local and federal regulations may require exhaust facilities if much lead soldering is done. Even if lead fumes are not present in harmful quantities, it is desirable to exhaust the nuisance fumes and smoke. Air samples should be taken to verify that the amount of lead in the air is not harmful.

Lead solder particles should not be allowed to accumulate on the floor and on work tables. If the operation is such that the solder or flux may spatter, employees should wear face shields or do the work under a transparent shield.

Air power tools

Air hose. An air hose presents the same tripping or stumbling hazard as do power cords on electric tools. Persons or material accidentally hitting the hose may unbalance the operator or cause the tool to fall from an overhead place. An air hose on the floor should be protected against trucks and pedestrians by two planks laid on either side of it or by a runway built over it. It is preferable to suspend hoses over aisles and work areas.

Workmen should be warned against disconnecting the air hose from the tool and using it for cleaning machines or removing dust from clothing.

Accidents sometimes occur when the air hose becomes disconnected and whips about. A short chain attached to the hose and to the tool housing will keep the hose from whipping about if the coupling should break. Air should be cut off before attempting to disconnect the air hose from the air line. Air pressure inside the line should be released before disconnecting.

A safety check valve installed in the air line at the manifold will shut off the air supply automatically if a fracture occurs anywhere in the line.

If kinking or excessive wear of the hose is a problem, it can be

Safety glasses saved the left eye of this plane overhaul mechanic. While he was using a pneumatic screwdriver to install the plate, the screwdriver tip broke and struck the left lens of his safety glasses. The lens was chipped, but the mechanic's eye was unharmed.

protected by a wrapping of strip metal or wire. One objection to armored hose is that it may become dented and thus restrict the flow of air.

Air power grinders require the same type of guarding as electric grinders. Maintenance of the speed regulator or governor on these machines is of particular importance in order to avoid over-speeding the wheel (runaway).

Regular inspection by qualified personnel at each wheel change is recommended.

Pneumatic impact tools, such as riveting guns and jackhammers, are essentially the same in that the tool proper is fitted into the gun and receives its impact from a rapidly moving reciprocating piston driven by compressed air at about 90 psig pressure.

Two safety devices are needed. The first is a trigger located inside the handle where it is reasonably safe from accidental operation. The machine operates only when the trigger is depressed. The second is a device to hold the tool in place so that it cannot be shot accidentally from the barrel.

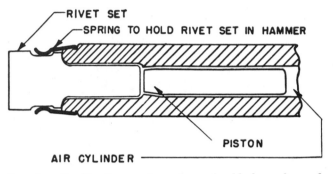

A spring clip, like the one shown here, should always be used to prevent a rivet set from falling from a hammer.

On small air hammers not designed to use this device, use a spring clip to prevent the tool and piston from falling from the hammer.

It is essential that employees be thoroughly trained in their proper use. A safety rule to impress on all operators of small air hammers is *Do not squeeze the trigger until the tool is on the work.*

Air-operated nailers and staplers. The principal hazard from these tools is the accidental discharge of the fastener. In such instances, the fastener can become a dangerous projectile and inflict serious injury at considerable distance. Operators should be trained in the use of these tools and must follow the manufacturer's operating instructions.

In the use of all pneumatic impact tools there is, of course, a hazard from flying chips. Operators should wear safety goggles, and, if other employees must be in the vicinity, they should be similarly protected. Where possible, screens should be set up to shield persons nearby where chippers, riveting guns, or air drills are being used.

Eye protection should be provided when using electric saws, grinding wheels, buffers, scratch brushes, and sanders.

Two chippers should work away from each other, that is, back to back, to prevent face cuts from flying chips. Workmen should not point a pneumatic hammer at anyone, nor should they stand in front of operators handling pneumatic hammers.

Handling of heavy jackhammers causes fatigue and may even cause strains. Jackhammer handles should be provided with heavy rubber grips to reduce vibration and fatigue, and operators should wear safety shoes to reduce the possibility of injury should the hammer fall.

Many accidents are caused by breaking of the steel drill because the operator loses his balance and falls. Also, if the steel is too hard, a particle of metal may break off and strike him. The manufacturer's instructions for sharpening and tempering the steel should be followed.

Special power tools

Flexible shaft tools require the same type of personal protective equipment as do direct power tools of the same type. Abrasive wheels should be installed and operated in conformance with American National Standard B7.1, as discussed earlier on pages 360 and 361. The flexible shaft must be protected against denting and kinking, which may damage the inner core and shaft.

It is important that the power be shut off whenever the tool is not in use. When the motor is being started, the tool end should be held with a firm grip to prevent injury from sudden whipping. The abrasive wheel or buffer of the tool is difficult to guard and, because it is more exposed than the wheel or buffer on a stationary grinder, extra care should be exercised to avoid damage. Wheels should be placed on the machine or put on a rack, not on the floor.

Gasoline power tools are widely used in logging, construction, and other heavy industry. Probably the best known and most prevalent is the chain saw. (See OSHA regulations § 1910.266(c)(5), chain saw operations, and NSC Data Sheet 320, *Portable Power Chain Saws.*)

Operators of gasoline power tools must be trained in their proper operation according to the manufacturer's instructions. They must also be familiar with the fuel hazards.

Explosive-actuated fastening tools (see OSHA regulations § 1910.243(d) and NSC Data Sheet 236, *Powder-Actuated Hand Tools*) are used for fastening fixtures and materials to metal, pre-cast, pre-stressed, concrete, masonry block, brick, stone, and wood surfaces, tightening rivets, and punching holes. Blank cartridges provide the

Powder-actuated tool drives threaded studs into poured concrete slab. Wearing of eye protection and safety hat is "standard operating procedure."

energy and are ignited by means of a conventional percussion primer.

The hazards encountered in the use of these tools are similar to those of a firearm. The handling, storing, and control of explosive cartridges present additional hazards. Therefore, instructions for the use, handling, and storage of both tools and cartridges should be just as rigid as those governing blasting caps and firearms.

Specific hazards are accidental discharge, ricochets, ignition of explosive or combustible atmospheres, projectiles penetrating the work,

Workman demonstrates correct use of power-assisted hammer-driven tool to fasten conduit clips to masonry blocks.

and flying dirt, scale, and other particles. In case of misfire, the operator should hold the tool in operating position for at least 30 seconds. He should try to fire the tool a second time. If it still misfires, he must hold the tool in operating position for another 30 seconds, then follow the manufacturer's instructions to remove the load.

Powder-actuated tools can be used safely if special training and proper supervision are provided for the operator. Manufacturers of the tools will aid in this training. Only trained and properly qualified personnel should be permitted to operate or handle the tools. A worker, however, may be qualified in a few hours of instruction.

Power-assisted, hammer-driven tools are used for the same purposes as powder-actuated tools and generally the same precautions should be followed.

Personnel should be trained to use powder charges of the correct size to drive studs into specific surfaces and should be made responsible for safe handling and storing of the cartridges and the tools.

Powder-actuated tools should not be used on concrete less than 2 in. thick nor on steel less than ¼ in. thick unless suitable material, such as timber or sandbags, that will contain the flying stud are placed directly behind the work. This material should be left in place until the job is completed. Fasteners should not be driven closer than 3 in. from an unsupported edge or corner. A low-velocity tool may drive no closer than 2 in. from an edge in concrete or ¼ in. in steel.

Operators should wear adequate eye protection when firing the tool. Where the standard shield cannot be used for a particular operation, special shields can be obtained from the tool manufacturer. Hearing protection should also be worn. See Chapter 7, Personal Protective Equipment, for more details.

CHAPTER 12

Fire Protection

A fire protection program must have the understanding and the cooperation of every supervisor in order to be effective. Although the overall program may be under the direction of the safety, fire protection, security, engineering, or maintenance staff, each foreman or supervisor on the job does have a direct interest in and responsibility for the program.

Fire protection, like other specialized fields, is a science in itself. This Manual cannot cover fire prevention and extinguishment in their entirety, but this chapter does present condensed, basic information to help the supervisor and his working group conduct fire-safe operations. More specialized information is available. Sources are cited throughout this chapter.

Basic Principles

The supervisor, because of his knowledge of the operations, is in an excellent position to determine the fire prevention measures necessary in his department. He should be able to recognize the need for specific fire protection equipment and should take the necessary steps to see that such equipment is provided. He should become thoroughly

It is apparent from this photo of a building construction fire that good housekeeping could probably have averted it.

familiar with the use of the fire equipment in his area.

Departmental housekeeping is also under the supervisor's control. He knows whether or not his employees follow safe housekeeping practices with regard to preventing fires.

Continuous training in fire-safe work procedures, regular inspections of work areas, and close supervision of the employees' job performance are requisites of a successful fire prevention program.

Although the fire protection equipment may be maintained by others, the ultimate responsibility for the fire-safety of employees, material in process, and production equipment rightfully belongs to supervision. Thus, every supervisor must make sure not only that the proper fire protection equipment is provided, but also that employees follow fire-safe work procedures.

Understanding fire chemistry

The supervisor's approach to fire prevention and control problems should begin with an understanding of basic fire chemistry. Every *ordinary* fire (one that does not produce its own oxygen supply) results when a substance (fuel) in the presence of air (oxygen) is heated to a critical temperature, called its "ignition temperature."

For many years the principle of extinguishment has centered around the "fire triangle" and the removal of any of its three sides, which are representative of three components. This principle does not include extinguishment by stopping the fire's chemical chain reaction, the fourth component.

The "fire pyramid" shows the four components necessary to produce ordinary burning. Remove any one, and the fire goes out.

Remodeling the "fire triangle" into a "fire pyramid" presents a more realistic concept of extinguishment. The pyramid has four sides or faces, one for each of the four ways to extinguish a fire. Because each face is directly adjacent to and connected to each of the other three, a pyramid accurately represents the interdependency that the three components and the chemical chain reaction have with respect to one another. The removal of one or more of the faces will make the pyramid incomplete and result in extinguishment.

Most commonly, fires are prevented by eliminating either the fuel or the heat (source of ignition). The best method of eliminating heat is to see that there are no ignition sources around combustibles.

Table XVI lists the sources of ignition that cause industrial fires, gives examples in each case, and suggests preventive measures. A following section, "Causes of Fire," discusses each in detail.

Large CC₂ extinguishers are used by these fire brigade members to extinguish a "practice fire."

Heat-actuated sprinkler system is timed to make sure it will operate fast enough and with enough water volume to control any fire on this tank farm.

372

To extinguish a fire, one or more of the following steps should be taken:

1. Exclude the air by smothering (for example, by shutting the lid over a tank of burning solvent or with foam) or by dilution (replacing the air with an inert gas such as carbon dioxide).
2. Remove or seal off the fuel by mechanical means, or divert or shut off the flow of burning liquids or gases.
3. Cool the burning material below its ignition point with a suitable cooling agent (hose streams or water extinguishers).
4. Interrupt the chemical chain reaction of the fire (using dry chemical extinguishing agents).

Once the supervisor understands and appreciates the fire pyramid and its practical application, he will be in a position to improve and extend his fire prevention and control methods.

Determining fire hazards

To best contribute to his organization's fire protection program, the supervisor must, first, determine the specific fire problems existing in his area of operations and, second, plan and take action to solve them. *He should take both steps with the help of the best technical advice he can get from experts.*

If he has not already done so, he should draw up an inspection checklist that will name as many places, materials, procedures, classes of equipment, conditions, and circumstances as possible—ones wherein fire hazards are likely to exist and which should be examined. Under such category names, used as headings on the list, specific fire-safe practices and installations should be cited concisely. Personnel with responsibilities relative to the practices and installations might be specified, as for example: machine operator, electrician, maintenance man, porter. (See discussion of inspections in Chapter 8.)

When, after a tour of inspection, the supervisor has made brief notations beside each of the described safe practices (or has filled in boxes with a "yes" or "no") to indicate observance of them or the lack of it, and has done the same with respect to described installations to show whether or not they have been provided and are being maintained, he will have supplied himself with a detailed picture of how many or how few fire protection measures are in effect in his depart-

Pressure in a sprinkler system riser is checked.

ment, or as relates to one or more particular jobs in it. He will be able to spot neglected precautions readily. The sample "Fire Prevention Checklist" shown here can serve as a guide, but the supervisor should make his own and include special points for factors and operations under his jurisdiction that are not listed in the broad example. A

FIRE PREVENTION CHECK LIST

ELECTRICAL EQUIPMENT

- ☐ No makeshift wiring
- ☐ Extension cords serviceable
- ☐ Motors and tools free of dirt and grease
- ☐ Lights clear of combustible materials
- ☐ Safest cleaning solvents used
- ☐ Fuse and control boxes clean and closed
- ☐ Circuits properly fused
- ☐ Equipment approved for use in hazardous areas (if required)
- ☐ Ground connections clean and tight

FRICTION

- ☐ Machinery properly lubricated
- ☐ Machinery properly adjusted and/or aligned

SPECIAL FIRE-HAZARD MATERIALS

- ☐ Storage of special flammables isolated
- ☐ Nonmetal stock free of tramp metal

WELDING AND CUTTING

- ☐ Area surveyed for fire safety
- ☐ Combustibles removed or covered
- ☐ Permit issued

OPEN FLAMES

- ☐ Kept away from spray rooms and booths
- ☐ Portable torches clear of flammable surfaces
- ☐ No gas leaks

PORTABLE HEATERS

- ☐ Set up with ample horizontal and overhead clearances
- ☐ Secured against tipping or upset
- ☐ Combustibles removed or covered
- ☐ Safely mounted on noncombustible surface
- ☐ Not used as rubbish burners

HOT SURFACES

- ☐ Hot pipes clear of combustible materials
- ☐ Ample clearance around boilers and furnaces
- ☐ Soldering irons kept off combustible surfaces
- ☐ Ashes in metal containers

SMOKING AND MATCHES

- ☐ "No smoking" and "smoking" areas clearly marked
- ☐ Butt containers available and serviceable
- ☐ No discarded smoking materials in prohibited areas

SPONTANEOUS IGNITION

- ☐ Flammable waste material in closed, metal containers
- ☐ Flammable waste material containers emptied frequently
- ☐ Piled material cool, dry, and well ventilated
- ☐ Trash receptacles emptied daily

STATIC ELECTRICITY

- ☐ Flammable liquid dispensing vessels grounded or bonded
- ☐ Moving machinery grounded
- ☐ Proper humidity maintained

HOUSEKEEPING

- ☐ No accumulations of rubbish
- ☐ Safe storage of flammables
- ☐ Passageways clear of obstacles
- ☐ Premises free of unnecessary combustible materials
- ☐ No leaks or drippings of flammables and floor free of spills
- ☐ Fire doors unblocked and operating freely with fusible links intact

EXTINGUISHING EQUIPMENT

- ☐ Proper type
- ☐ In proper location
- ☐ Unobstructed
- ☐ Clearly marked
- ☐ In working order
- ☐ Service date current
- ☐ Personnel trained in use of equipment

Sample check list serves as guide for supervisor in drawing up inspection list for his own work area. List should be reviewed regularly to keep it up-to-date.

fuller discussion is given in the next section, "Causes of fire."

The supervisor, when conducting an inspection within his department or of one particular job, may be amazed at the detailed points to be observed, some of which may have escaped his notice previously. Unless a man has had considerable experience in fire prevention, he should ask his superior and the company's fire safety personnel or the fire insurance company engineer to help him conduct inspections and to make recommendations for eliminating fire hazards. If company policy permits, the supervisor usually may ask the local fire department for help, too.

It is important that all fire inspections be made with a critical eye. Every shortcoming should be listed. The supervisor should not hedge on listing certain hazards with the thought that they might reflect poor supervision. Omission of a pertinent detail might result in a fire later; such an outcome *would* reflect poor supervision.

The NFPA *Inspection Manual,* a pocket-sized book, is a valuable reference for the beginner as well as the experienced inspector in setting up and conducting inspections. Its content covers both common and special fire hazards, their elimination or safeguarding, building construction, and human safety in all types of properties.

Informing the working force

Periodic inspections by the supervisor are an important part of his department's fire protection program. However, the supervisor's responsibilities, under a complete program, extend further. As he becomes acquainted with actual or potential fire hazards, and after all physical corrections possible have been made, he should familiarize the men in his department with each hazard and explain how it relates to them individually.

He should inform his men of his and management's desire for fire-safe operations. He should call their attention to all the physical safeguards that have been provided to prevent injury and destruction by fire. He should stress to each individual the precautions necessary on *his* job—the safe practices that complement mechanical protection.

If the man on the job understands the reason for the rule and the possible consequences if he does not follow it, he is much more apt to comply. Patient explanation and persistent enforcement, in every case, are two prime fire prevention duties of the supervisor.

Causes of Fire

The supervisor should be alert for potential causes of fire. The principal ones shown in Table XVI are discussed.

Electric equipment

Electric motors, switches, lights, and other electrical equipment exposed to flammable vapors, dusts, gases, or fibers present special problems. The NFPA's "National Fire Codes" designate the standard governing a particular hazard and indicate the special protective equipment needed. The *National Electrical Code*, NFPA Std. No. 70, gives the specifications for the protective equipment required. Substandard substitutions or replacements must not be made.

Haphazard wiring, poor connections, and "temporary" repairs must be brought up to standard. Fuses should be the proper type and size. Circuit breakers should be checked to see that they have not been blocked in the closed position, which results in overloading, and to see that moving parts do not stick.

Cleaning electric equipment with solvents can be hazardous since many solvents available for this purpose are both flammable and toxic. It is, of course, of utmost importance to use the safest cleaning solvents available.

A solvent may be safe with respect to fire hazards but very unsafe with respect to health hazards. For example, carbon tetrachloride is nonflammable, but its vapors are extremely toxic. Before a solvent is used, therefore, it is necessary to determine both its toxic properties and its flammable properties.

A satisfactory solvent is inhibited methyl chloroform, or a blend of Stoddard solvent and perchloroethylene. These solvents are commonly used in industry since they are relatively nonflammable and have a relatively high threshold limit value with respect to toxicity. (See Chapter 6 for a discussion.)

Many of the cleaning solvents encountered in industry are not single substances, but mixtures of different chemicals, usually marketed under nondescriptive trade names or code numbers. Currently, there are no absolutely safe cleaning solvents. Therefore, before any commercially available solvent is used, it is essential to know its chemical composition. Without such knowledge, the hazards cannot be evaluated nor the required safety controls be used. (Again, see Chapter 6.)

Friction

Overheated transmission bearings and shafting where dust and lint accumulate, as in grain elevators, cereal and textile mills, and plastic, woodworking, and metalworking plants, are frequent sources of ignition. Bearings should be kept lubricated so that they do not run hot, and accumulations of flammable dust on them should be removed as part of a rigid housekeeping routine. Pressure lubrication fittings should be kept in place and oil holes of bearings should be kept covered to prevent combustible dust and grit from entering the bearings and causing overheating.

Special fire-hazard materials

Certain materials must be kept isolated to prevent fire. For example, some chemicals, like sodium and potassium, decompose violently in the presence of water, evolve hydrogen, and ignite spontaneously.

Yellow phosphorus may ignite spontaneously on exposure to air. Other combinations, too numerous to mention here, may react with the evolution of heat and produce fire or explosion—in some cases, without air or oxygen being present. Such materials must be handled in a special manner.*

Also, some materials, principally the ethers, during long periods of storage may become unstable and eventually explosive. In such cases, using the oldest stock first contributes to both fire safety and good housekeeping—and seeing that this principle is followed is part of the supervisor's job.

Whatever materials are used or stored in the department, it is important to know whether they explode when heated, react with

* "National Fire Codes," Vol. I, *Flammable Liquids,* Vol. II, *Gases,* and Vol. III, *Combustible Solids, Dusts, and Explosives,* published by the National Fire Protection Association, Boston, are excellent sources of detailed information on this subject. Vol. III includes NFPA Standard No. 49, *Hazardous Chemicals Data,* which lists about 100 items. The table includes information on usual shipping containers, fire hazards, life hazards, storage, fire fighting phases, and additional data, where applicable.

TABLE XVI. SOURCES OF IGNITION IN INDUSTRIAL FIRES

SOURCES OF IGNITION (Descending order of frequency)	EXAMPLES	PREVENTIVE MEASURES
Electrical equipment	Electrical defects, generally due to poor maintenance, mostly in wiring, motors, switches, lamps, and hot elements.	Use only approved equipment. Follow *National Electrical Code*. Establish regular maintenance.
Friction	Hot bearings, misaligned or broken machine parts, choking or jamming of material, poor adjustment.	Follow a regular schedule of inspection, maintenance, and lubrication.
Foreign substances	Tramp metal, which produces sparks when struck by rapidly revolving machinery (a common cause in textile industry).	Keep foreign material from stock. Use magnetic or other separators to remove tramp metal.
Open flames	Cutting and welding torches (chief offenders). Gas and oil burners. Misuse of gasoline torches.	Follow established welding precautions. Keep burners clean and properly adjusted. Do not use open flames near combustibles.
Smoking and matches	Dangerous near flammable liquids and in areas where combustibles are used or stored.	Smoke only in permitted areas. Use prescribed receptacles. Make sure matches are out.
Spontaneous ignition	Deposits in ducts and flues. Low-grade storage. Industrial wastes. Oily waste and rubbish.	Clean ducts and flues frequently. Remove waste daily. Isolate stored materials likely to heat spontaneously.
Hot surfaces	Exposure of combustibles to furnaces, hot ducts or flues, electric lamps or irons, hot metal being processed.	Provide ample clearances, insulation, air circulation. Check heating apparatus before leaving it unattended.
Combustion sparks	Rubbish-burning, foundry cupolas, furnaces and fireboxes, and process equipment.	Use incinerators of approved design. Provide spark arresters on stacks. Operate equipment carefully.
Overheated materials	Abnormal process temperatures. Materials in driers. Overheating of flammable liquids.	Have careful supervision and competent operators, supplemented by well-maintained automatic temperature controls.
Static electricity	Dangerous in presence of flammable vapors. Occurs at spreading and coating rolls or where liquid flows from pipes.	Ground equipment. Use static eliminators. Humidify the atmosphere.

Adapted from *Factory Mutual Record*

water, heat spontaneously, yield hazardous decomposition products, or otherwise react in combination with other materials.

The supervisor's best course of action is to obtain detailed information and be guided accordingly. The company's fire insurance carrier can be asked for help.

Welding and cutting

If at all possible, welding and cutting operations should be done in a separate, well-ventilated room with a fire-resistant floor. This safety measure is not, of course, always practical.

If welding and cutting must be carried on in other locations, these operations must not be performed until (a) the areas have been surveyed for fire safety by persons who know the hazards, (b) the necessary precautions to prevent fires have been taken, and (c) a permit has been issued. This permit must not be "stretched" to cover an area or an item or a time not originally specified—no matter how small the job may seem nor how little time may be required to do it.

If welding must be done over wood floors, they should be swept clean, wet down, and then covered with asbestos blankets, metal, or other noncombustible covering. Pieces of hot metal and sparks must be kept from falling through floor openings onto combustible materials.

Sheet metal, flame-resistant canvas, or asbestos curtains should be used around welding operations to keep sparks from reaching combustible materials. Welding or cutting should not be permitted in or near rooms containing flammable liquids, vapor, or dust. Neither should it be done in or near closed tanks which contain—or have contained—flammable liquids until the tanks have been thoroughly drained and purged and tested free from explosive gases or vapors.

No welding or cutting should be done on a surface until combustible coverings or deposits have been removed. It is important that flammable dusts or vapors not be created during the welding operation. Fire extinguishing equipment should be provided at each welding or cutting operation as standby equipment. A water pump tank unit is recommended for Class A fires, and a dry chemical or carbon dioxide for Class B and C fires, discussed later in this chapter. A watcher should be stationed to prevent stray sparks or slag from starting fires or to immediately extinguish fires that do start while they are small. The area should be under fire surveillance for at least one-half hour *after* welding or cutting has been completed since many fires are not detected as soon as they start.

Open flames

There must be no open flames in or near spray rooms or spray booths. Occasionally, it may be necessary to do indoor spray-painting or spray-cleaning outside of a standard spray room or booth, which is provided with electrical equipment that is listed for the hazardous location, and a proper ventilating system. In such cases, adequate ventilation must be provided and possible ignition sources, such as spark-producing devices and open flames, must be eliminated.

Gasoline, kerosene, or alcohol torches should be placed and used so that their flames are at least 18 in. from wood surfaces. They should not be used in the presence of dusts or vapors, or near flammable liquids, paper, excelsior, or similar material. Torches should not be left unattended while they are burning.

Portable heaters

Gasoline furnaces, portable heaters, and salamanders present a serious fire hazard. Their use should be discouraged as much as possible.

Fuel used in salamanders and other portable heaters should be restricted to liquefied petroleum gas, coal, coke, fuel oil, or kerosene. The area in which they are burned must be well ventilated since, like other heaters, they can produce carbon monoxide.

All these heating devices require attention with respect to clearances and mounting. A clearance of 2 ft horizontally and 6 ft vertically should be maintained between a heater and any combustible material. If coal or coke is used, the heater should be supported on legs 6 in. high or on 4 in. of tile blocks, and set on a noncombustible surface.

Flammable material overhead should be removed or shielded by noncombustible insulating board or sheet metal with an air space between it and the combustible material. A natural-draft hood and flue of noncombustible material should be installed.

As a fire precaution, each unit must be carefully watched. Fuel oil and kerosene salamanders should be shut down before they are moved or refueled. They should be allowed to cool off before being refueled. Coal and coke salamanders should not be moved until the fire is out.

All portable heating devices should be equipped with suitable handles for safe and easy carrying. They also should be secured or protected against tipping or upsetting.

Of the portable heating devices mentioned, the salamander presents the most serious fire hazard if it is not properly handled. Salamanders are too often improvised of old steel drums or empty paint containers, with scrap wood, tar paper, or other waste used as fuel. This practice should be discouraged since it usually permits little control of sparks and smoke.

Rubbish should be burned in rubbish burners that have wire mesh-screen covers. The burners should be placed at least 50 ft from combustible stores and equipment. In view of recent anti-pollution laws, make sure that no violation of clean air ordinances are being committed.

Hot surfaces

If possible, smoke pipes from heating appliances should not pass through ceilings or floors. If a smoke pipe must be run through a combustible wall, a galvanized double-thimble with clearance equal to the diameter of the pipe and ventilated on both sides of the wall must be provided.

Soldering irons must not be placed directly upon wood benches or other combustibles. Rests, which will prevent dangerous heat transfer, can be used to keep them off such surfaces.

Smoking and matches

Management usually has a specific policy with regard to smoking by employees. "Smoking" areas as well as "no smoking" areas must be clearly defined and marked off with conspicuous signs. Reasons for these restrictions must be clearly explained to the employees, and rigid enforcement must be maintained *all* the time with no exceptions.

Fire-safe, metal butt containers should be provided in places where smoking is permitted. If the carrying of matches is prohibited, special lighter equipment should be kept in service in smoking areas.

"No smoking" areas, especially when they include stairways and other out-of-the-way places, should be watched for evidence of discarded smoking materials.

Spontaneous ignition

Spontaneous ignition is a chemical action in which there is a slow generation of heat from the oxidation of a fuel until the ignition temperature of the fuel is reached. The fuel then begins to burn. Conditions leading to spontaneous ignition exist where there is sufficient air for oxidation but not enough ventilation to carry away the heat as fast as it is generated. Any factor that accelerates the oxidation while other conditions remain constant obviously increases the likelihood of such ignition.

Materials like unslaked lime and sodium chlorate are susceptible to spontaneous ignition, especially when wet. Such chemicals should be kept cool and dry, away from combustible material. Rags and waste saturated with linseed oil or paint often cause fires because no provision is made for the generated heat to escape. By keeping such refuse in air-tight metal containers with self-closing covers, the oxygen supply is limited and a fire will quickly extinguish itself. The containers should be emptied daily.

The best precaution against spontaneous ignition is either total exclusion of air or good ventilation. The former is practicable with small quantities of material through the use of air-tight containers. Ventilation can best be assured by storing material in small piles or by "turning" a large pile at regular intervals.

To determine the progress of spontaneous heating, temperatures should be taken in the interior of a mass of material. Various locations within the mass should be checked. Exterior temperatures are not likely to provide a good index to the progress of the heating.

Static electricity

Sparks due to static electricity may be a hazard wherever there are flammable vapors or gases or combustible dusts. Precautions against static electricity are required in such areas. Static charges result from friction between small particles or from the contact and separation of two unlike substances, one or both of which are nonconductive.

Static charges can be produced in many ways. An example is the

Air tight, sheet metal waste can. Cover cannot be opened more than 60 degrees so that it will close automatically.

Automatic take-up reel holds bonding wire for safety cans while they are being filled. Drum rack is grounded.

blue continuous spark which can be seen when friction tape is unrolled in the dark. Static charges may also be produced by the flow of flammable liquids through a nonconductive hose or by the passing of dry and powdered materials down a nonconductive chute or through a machine.

It is impossible to prevent the generation of static electricity under these circumstances, but the hazard of static sparks can be avoided by preventing the accumulation of static charges. Any one or more of the following methods may be used:

• Grounding,
• Bonding,
• Maintaining the relative humidity at a predetermined level, or
• Ionization of the atmosphere.

A combination of these methods may be advisable in some instances where the accumulation of static charges presents a severe

Grounds and bonds should be constructed of bare flexible wire; this facilitates inspection and prevents broken wires from being concealed.

hazard. The National Fire Protection Association's Standard No. 77, *Static Electricity*, gives additional information and guidance on this subject.

Grounding is accomplished by mechanically connecting a conductive machine or vessel in which the generation of static may be a hazard to ground by means of a conductor of low resistance. Another method is to make the entire floor and structure of the building conductive so that all equipment in contact with it will be grounded. When the former method is used, the supervisor must check the continuity of the ground circuit. Connections must be clean and the conductor unbroken. With the latter method, it is important that the floor be free of wax, oil, or other insulating films.

As many people have learned from walking across a rug and then touching a door knob or other conductive object, the human body can carry an electrical charge. The use of conductive shoes with a floor of conductive material is a common means of controlling this hazard by the grounding method. Parts of such shoes that are designed to render them conductive should be made of nonferrous metal. As an added measure of safety, the conductive flooring may be made of spark-resistant metal. Ferrous contacts increase the risk of friction sparks.

When humidity is low, the hazard of static is greatest. When humidity is high, the moisture content of the air serves as a conductor to drain off static charges as they are formed. Where humidification is utilized to prevent the accumulation of static charges, the supervisor must see that an effective relative humidity—usually 60 to 70 percent—is maintained. However, the minimum humidity required for safety may vary over a considerable range under different conditions, and under some conditions the static charge cannot be controlled by humidification. Engineering authorities should be consulted.

When air is ionized, it has sufficient conductivity to prevent static accumulation. Ionization is produced by electrical discharges, radiation from radioactive substances, or gas flames. Only an electrostatic neutralizer designed for use in hazardous locations should be used; otherwise, the neutralizer may itself be a source of ignition of flammable vapor or dust. It must be kept in good condition.

Fire-safe Housekeeping

Good housekeeping is another important part of an effective fire protection program. It is imperative that the supervisor maintain strict discipline with his crew and enforce the housekeeping rules at *all* times. Each man should be held personally responsible for preventing the accumulation of unnecessary combustible materials in his work area. Especially, he should be held accountable for his area at the end of his shift. Here are the precautions to take.

1. Combustible materials should be present in work areas only in quantities required for the job, and should be removed to a designated, safe storage area at the end of each work day.

2. Quick-burning and flammable materials should be stored only in designated locations. Such locations always should be away from

ignition sources and have special fire extinguishing provisions.

3. Vessels or pipes containing flammable liquids or gases must have no leaks. Any spills should be cleaned up immediately.

4. Workmen should be required to guard carefully against any part of their clothing becoming contaminated with flammable liquids. If contamination does occur, they must be required to change such clothing before continuing to work.

5. Passageways and fire doors should be kept clear and unobstructed.

6. Material must not obstruct sprinkler heads or be piled around fire extinguisher locations or sprinkler and standpipe controls. To obtain proper distribution of water, a minimum of 18 in. of clear space is required below sprinkler deflectors. However, clearance of 24 to 36 in. is recommended. If there are no sprinklers, clearance of 3 ft between piled material and the ceiling is required to permit room for the use of hose streams. Double these distances when stock is piled more than 15 ft high. Be sure to check applicable codes.

Alarms, Equipment, and Evacuation

A complete and fully effective fire program, of course, includes fire extinguishment. Despite the best laid preventive plans, a fire may occur. Whenever a fire occurs, it is important to:

1. Turn in the alarm right away—regardless of the size of the fire.

2. Attempt to extinguish or control the fire with appropriate fire extinguishing equipment promptly—in order to minimize hazard.

Fire alarms

Plants in areas where municipal fire departments are available usually have a municipal alarm box close to the plant entrance or located in one of their buildings. Others may have auxiliary alarm boxes, connected to the municipal fire alarm system, at various points in the plant. Another system often used is a direct connection to the nearest fire station which may register by a water alarm on the sprinkler system or be set off manually. In some cases, the telephone may be the means for signaling a fire alarm.

Whatever the alarm system used, *all* employees should be given careful instruction in (*a*) how to report a fire, (*b*) when, and (*c*) where. These three items are extremely important since many fires progress beyond control simply because someone did not know how or when or where to give the alarm.

Alarm boxes and fire fighting equipment must be clearly marked and free of obstructions. Fire doors must be kept clear.

What about extinguishers?

The foreman should know what class or classes of fires might be expected in his department. Before the employees, working under the direction of their supervisor, can effectively combat fires in their incipient stages, the classes of fires must be understood. Briefly, they are:

Class A—Fires in ordinary combustible materials, such as wood, paper, or clothing, where the quenching and cooling effects of water or of solutions containing large percentages of water are of prime importance.

Class B—Fires in flammable liquids, greases, and similar materials, where smotherings or exclusion of air, and interrupting the chemical chain reaction is most effective.

Class C—Fires in or near live electric equipment where the use of a nonconductive extinguishing agent is of first importance. The material that is burning is, however, either Class A or Class B in nature.

Class D—Fires that occur in combustible metals such as magnesium, lithium, and sodium. Special extinguishing agents and techniques are needed for fires of this type.

Recommended colors as described in the Federal Color Standard Number 595 are:

Class A—Green, No. 14260

Class B—Red, No. 11105

Class C—Blue, No. 15102

Class D—Yellow, No. 13655

Each of the fire extinguishers in the department or at a job location should have on it a data plate giving the class of fire for which it is intended, operating instructions, and servicing instructions. The data plate should carry the label of either Underwriters' Laboratories, Inc., or Factory Mutual, to indicate that the unit has been listed or approved. Equipment which does not bear either label should be brought to the attention of management. Only listed or approved

FIRE EXTINGUISHER/AGENT

SUITABLE FOR USE ON TYPE OF FIRE	AGENT CHARACTERISTICS	Available Sizes	Horizontal Range	Discharge Time
REGULAR OR ORDINARY DRY CHEMICAL ★				
B **C**	Basically Sodium Bicarbonate. Discharges a white cloud. Leaves residue. Non-freezing.	1 to 30 lbs.	5 to 20 ft.	8 to 25 Sec.
MULTIPURPOSE DRY CHEMICAL ★				
A **B** **C** OR **B** **C** **A** CAPABILITY	Basically Ammonium Phosphate. Discharges a yellow cloud. Leaves residue. Non-freezing. Some extinguishers utilizing this agent do not have an "A" rating — however, they are designated as having "A" capability.	2 to 30 lbs.	5 to 20 ft.	8 to 25 Sec.
PURPLE-K DRY CHEMICAL ★				
B **C**	Basically Potassium Bicarbonate. Discharges a bluish cloud. Leaves residue. Non-freezing.	2 to 30 lbs.	5 to 20 ft.	8 to 25 Sec.
KCL DRY CHEMICAL ★				
B **C**	Basically Potassium Chloride. Discharges a white cloud. Leaves residue. Non-freezing.	2 to 30 lbs.	5 to 20 ft.	8 to 25 Sec.
	Potassium Chloride/Urea	11 to 23	15 to 30	20 to 31
CARBON DIOXIDE				
B **C**	Basically an inert gas that discharges a cold white cloud. Leaves no residue. Non-freezing.	2½ to 20 lbs.	3 to 8 ft.	8 to 30 Sec.

★ NOTE: Available in stored pressure or cartridge operated types.

390

CHARACTERISTICS

SUITABLE FOR USE ON TYPE OF FIRE	AGENT CHARACTERISTICS	Available Sizes	Horizontal Range	Discharge Time
HALOGENATED AGENT B C	Basically halogenated hydro-carbons. Discharges a white vapor. Leaves no residue. Non-freezing.	2½ lb.	4 to 8 ft.	.8 to 10 Sec.
WATER ▲ A	Basically tap water. Discharges in a solid or spray stream. (May contain corrosion in-hibitor which leaves a yellow residue.) Protect from freezing!	2½ Gal.	30 to 40 ft.	1 Minute
ANTI-FREEZE SOLUTION A	Basically a Calcium Chloride solution to prevent freezing. Discharges a solid or spray stream. Leaves residue. Non-freezing.	2½ Gal.	30 to 40 ft.	1 Minute
LOADED STREAM A B	Basically an alkali-metal-salt solution to prevent freezing. Discharges a solid or spray stream. Leaves residue. Non-freezing.	2½ Gal.	30 to 40 ft.	1 Minute
FOAM B	Basically water and detergent. Discharges a foamy solution. After evaporation, leaves a powder residue. Protect from freezing.	18 oz.	10 to 15 ft.	24 Sec.
DRY POWDER SPECIAL COMPOUND D	Basically Sodium Chloride or Graphite materials. Agent is discharged from an extin-guisher in a solid stream or is applied with a scoop or shovel to smother combustible metal. Leaves residue. Non-freezing.	30 lbs.	5 to 20 ft.	25 to 30 Sec.

▲ NOTE: Pump tanks available.

Courtesy Fire Equipment Manufacturers' Association, Inc., Mt. Prospect, Ill. 60056.

In-plant, electric powered fire truck travels at 16 mph, and carries 850 ft of 2½-in. rubber-lined hose, one 150-lb and one 20-lb dry chemical extinguisher, two 2½-gal water extinguishers, one 20-lb CO_2 extinguisher, one self-contained breathing apparatus, a two-section 16-ft extension ladder, fire axe, and wrecking bar. An equipment cabinet houses wrenches, spanners, salvage covers, first aid kit, rope, and nozzles.

equipment should be furnished. *

The information on these data plates can be used by the supervisor in teaching employees how to operate the extinguishing equipment. The supervisor must see that every worker under his charge knows the important details about each of the fire extinguishing agents provided for use in his particular job area or shop.

Normally, the location and installation of portable fire extinguishers, fire pails, fire blankets, stretchers, and other fire-fighting equipment at strategic places about the shop or job site are the responsibility of higher management. However, the supervisor should know the location of units in his area, and should be quick to recommend reloca-

* "Fire Protection Equipment List" (Underwriters' Laboratories, Inc., Chicago 60611). "Approved Equipment for Industrial Fire Protection" (Factory Mutual Engineering Division, Norwood, Mass. 02062).

tion of them or procurance of additional units when he believes such changes will afford more adequate protection.

Mounting heights are spelled out by OSHA requirements. "Extinguishers having a gross weight not exceeding 40 lb shall be installed so that the top of the extinguisher is not more than 5 ft above the floor. Extinguishers having a gross weight greater than 40 lb (except wheeled types) shall be so installed that the top of the extinguisher is not more than 3½ ft above the floor."

Extinguishers must not be blocked by material or equipment, and that signs indicating their whereabouts remain legible and conspicuous. Many companies have found that painting the area of the floor directly under fire equipment is an excellent way to prevent employees from placing obstructions in front of the equipment.

The location of extinguishers can be identified by painting the housing, wall area, column, or other support of the extinguisher with standard fire protection red. Fire protection equipmerit itself, such as sprinkler systems, is often painted red. (See page 221.)

The supervisor should make sure that each of his workmen knows the location of the nearest unit and that each individual is impressed with the importance of cooperating to keep areas around extinguisher units clear.

It is advisable for the supervisor to have his men engage in drills in which they use extinguishers applicable to their particular work areas. The company's safety engineer or fire chief, the insurance company's safety engineer, or the local fire department representative will assist in setting up and conducting such drills.

Every organization should have a specific program for periodic inspection and servicing of portable fire extinguishing equipment. This routine work is usually outside the scope of the department foreman or supervisor. However, he can do some things to assist in this program. For example, during regular department inspections, he should double-check each data card to determine the date when each extinguisher was last serviced. By doing so, he may catch an omission which might make the difference between a small blaze and an inferno.

Followup for fire safety

Frequent inspections of the area, correction of hazardous conditions, and indoctrination of the workers in fire prevention and control measures are still not enough to assure a constantly fire-safe work area. The supervisor must follow up persistently. Constant alertness to assure a continuous fire-safe attitude on the part of personnel and their observance of safe work practices is required of the foreman.

It is recommended that fire prevention and control be the subject of frequent safety talks between the supervisor and each worker. Fire prevention and control should also be discussed in safety meetings so that each worker will be fully impressed that it is a definite part of the overall department safety program.

Fire brigades

Many plants and construction job sites have organized fire brigades. The organization of such a brigade would not normally be the responsibility of the supervisor. However, he should be well enough acquainted with the form and activities of a fire brigade to carry out the responsibilities he is likely to have in the brigade's operations.

In some fire brigades, a supervisor or foreman is designated a brigade chief or company captain. Whether or not the supervisor is named to this job, he should know the fire brigade assignments for which the men in his department are responsible.

On the basis of this information, the supervisor should then organize the work in his department so that brigade members may attend brigade training and drills as designated by their brigade chief.

Regardless of his role in fire brigade organization, the supervisor should be familiar with the location and operation of the following items in and adjacent to his department: standpipes and valves, sprinkler system valves, water pumps and controls, flammable gas and fluid lines and valves, electric switches and fans, alarm boxes and telephones, fire doors.

Enclosed stairway doors are especially important. For life safety, they *must* be kept closed, and it is a direct and serious responsibility of the supervisor to see that they *are* kept closed. Such doors are provided to keep smoke, gases, and heat out of the exit passageways and stairs, as well as to help prevent the rapid spread of fire.

To gain familiarity with the items listed above, the supervisor should take every opportunity to accompany the fire inspector on inspection tours of his department and to learn how the items operate.

Whether the supervisor himself should operate this equipment or

Plant fire protection diagram used in training fire brigade. Map shows sprinkler lines, pump houses, underground water supply systems, outdoor hose cabinets, fire hydrants, post indicator valves, and fire alarm and sprinkler alarm systems.

not will depend upon the particular situation and specific policy and procedure in the plant. In some instances, it may be desirable for the supervisor to operate this equipment if he knows the system thoroughly. However, some cases in the records of the National Fire Protection Association indicate that it may be extremely dangerous for anyone to operate the equipment without full knowledge concerning it. Operation at the wrong time or inaccurate operation might seriously affect the protective measures carefully set up by those in charge of fire protection in the plant.

Special fire protection problems

When building construction is going on at an existing plant site, some of the fire protection problems are compounded. The plant department supervisor and the construction supervisor should work together in determining the hazards in each field of operations and in maintaining awareness of fire fighting limitations. Definite fire preventive steps should be taken.

Such practices as use of temporary wiring and of portable heaters and gasoline engines that have to be refueled on the spot are likely to be required in the construction activities. Manufacturing processes, normally not hazardous, may become so under these conditions. The supervisor for the plant department should see that his men are aware of additional ignition sources created by the construction operations. The construction supervisor should inform his men of the additional exposures present from the plant activity.

When the construction work involves plumbing, the water supply available for fire extinguishment may be decreased. To offset this possible storage of water, additional auxiliary fire extinguishing equipment should be available. Either temporary hose lines or water tank trucks can be used.

Some plant and construction supervisors have made a practice of attending each other's safety meetings in order to benefit from an exchange of their respective points of view and to learn each other's problems. This joint approach to common exposures is recommended.

Another special fire protection problem is encountered whenever the sprinkler system must be shut down for alterations or repairs, because a serious weakness in fire protection results. The supervisor and the maintenance people, therefore, should do their best to plan such operations so that they can be handled outside normal working hours. If work must be done during the interval of danger, special precautions may have to be taken, such as having hose lines laid and furnishing extra patrols.

The use of radioactive materials for various purposes in manufacturing processes is becoming more common, and their presence requires special understanding and control in the event of fire or other emergency.

In view of the potential contamination hazards to employees, supervisors in the areas or departments having radioactive materials should be thoroughly familiar with the procedure to be followed in such areas or departments in case of fire or other emergency, and should rigidly enforce compliance with that procedure. Access to the contaminated area following the emergency should be prohibited until authorized personnel have recovered the sources of radiation and have

General emergency and/or fire alarm system is part of industrial security system.

determined that the contamination is below the safe levels.

Valuable information on this problem can be secured from the Atomic Energy Commission, Office of Technical Services, Washington, D.C. 20545.

Evacuation

Prevention of fire is the primary objective of any fire protection program. Nevertheless, each program must include provisions to assure the safety of employees in the event of a fire. Essential among such provisions are those that will chart and facilitate the quick and orderly evacuation of personnel if a serious fire emergency occurs.

The supervisor can do much to prepare for the safety of employees in his department in the event of a serious fire. As a matter of current routine, he should make sure that each man and woman knows the evacuation alarms, *both* the primary and alternate exits and escape routes, and what to do during and after an evacuation.

The supervisor should make sure that each person knows that he

397

or she, upon being alerted, must proceed at a fast walking pace—not a run—to an assigned exit or, if it is blocked, to the nearest clear one. Day in and day out, he should emphasize that exit routes through the department and fire doors must be kept unblocked. Furthermore, the supervisor should make it understood that possible damage and additional hazards can be avoided if each employee, upon a signal for evacuation, shuts down his or her machine before leaving it.

Such essential knowledge and habits to conform to emergency procedures will best be instilled in personnel by periodic fire drills. If company policy assigns the responsibility to the supervisor, he should plan and conduct practice evacuations at regular intervals within his area of operations, after availing himself of expert advice and technical assistance. He should also do everything possible to integrate fire emergency training in his department with plant-wide drills and the overall evacuation plan. When it is impossible to hold drills, the supervisor should give oral instructions regarding evacuation conduct, and should see that printed information is distributed periodically to employees.*

Reviewing the Supervisor's 'Fire Job'

The supervisor's prime responsibility with regard to fire prevention and control in his department is summarized in these general points:

1. Know the fire hazards in the department, and promote installation of engineered safeguards in every instance practicable.
2. Conduct regular, periodic inspections of work areas to assure that they remain in fire-safe condition.
3. Instill in each employe a fire-conscious attitude for elimination of hazards and observance of safe practices—then *follow up* his performance.
4. Keep well informed on fire brigade activity and evacuation plans. Keep employees apprised of any changes in procedure.

A foreman or supervisor will rightfully feel that the prime reason for his being in charge of a group of workers is to get out production—to get a job done safely. A vital element of that overall job is fire prevention and control.

* Details are given in NSC Data Sheet No. 588, *Fire Brigades*.

USEFUL REFERENCES

American National Red Cross, 17th and D Sts. NW., Washington D.C. 20006. *First Aid Textbook.*

American National Standards Institute, 1430 Broadway, New York, N.Y. 10018. "American National Standards" (catalog available).

American Welding Society, 2501 NW. 7th St., Miami, Fla. 33125. *Welding Handbook.*

Factory Mutual Engineering Corporation of the Factory Mutual System, 1151 Boston-Providence Turnpike, Norwood, Mass. 02062.
Approval Guide.
Handbook of Property Conservation.
Loss Prevention Data.

Firenze, Robert J. *Guide to Occupational Safety and Health Management.* Dubuque, Iowa, Kendall/Hunt Publishing Co., 1973.

Hannaford, Earle S. *Supervisors Guide to Human Relations.* Chicago, Ill. National Safety Council, 1967.

Illuminating Engineering Society, 345 East 47th St., New York, N.Y. 10017.
IES Lighting Handbook (The Standard Lighting Guide).
Practice for Industrial Lighting (ANSI A11.1–1965).

International Labor Organization, 666 11th St. NW., Washington, D.C. 20001. *Encyclopedia of Occupational Health and Safety,* 1972. (Available from McGraw-Hill Book Co., New York, N.Y. 10036.)

Matwes, George and Helen. *Loss Control: A Safety Guidebook for Trades and Services.* New York, N.Y., Van Nostrand Reinhold Co., 1973.

National Association of Suggestion Systems, 435 N. Michigan Ave., Chicago, Ill. 60611. *Journal* (quarterly).

National Fire Protection Association, 470 Atlantic Ave., Boston, Mass. 02110.
Fire Prevention Handbook.
Fire Protection Guide on Hazardous Materials.
Inspection Manual.
National Electrical Code, Std. No. 70 (ANSI C1–1971).
"National Fire Codes" (10 volumes).
"Standards and Recommended Practices" (catalog available).

399

Useful References

National Institute of Occupational Safety and Health, U.S. Department of Health, Education, and Welfare, 5600 Fishers Lane, Rockville, Md. 20852.
    "Criteria Documents."
    Toxic Substances List.
National Safety Council, 425 N. Michigan Ave., Chicago, Ill. 60611.
    Accident Facts (annually).
    Accident Prevention Manual for Industrial Operations.
    Basics of Industrial Noise and Audiometry.
    "Catalog and Poster Directory."
    Family Safety Magazine (quarterly)
    "5 Minute Safety Talks" (series).
    Fundamentals of Industrial Hygiene.
    "Guards Illustrated."
    How To Give the Safety Speech.
    "Industrial Data Sheets" (alphabetical listing available).
    National Safety News Magazine (monthly)
    OSHA Standards Checklists, Parts 1910 and 1926.
    "Pocket Guide to First Aid."
    "Power Press Safety Manual."
    "Safety Guide for Health Care Institutions."
    "Safety Slides" (a series).
    Successful Supervision.
    Supervisors Guide to Human Relations.
National Society for the Prevention of Blindness, Inc., 79 Madison Ave., New York, N.Y. 10016. *Eyesight in Industry.*
Underwriters' Laboratories, Inc., 207 E. Ohio St., Chicago, Ill. 60611. "Product Directories."
U.S. Department of Interior, Bureau of Mines, Washington, D.C. 20240. *Manual of First Aid Instructions.*
U.S. Department of Labor, Occupational Safety and Health Administration, Washington, D.C. 20210.
    Code of Federal Regulations. Title 10—Atomic Energy.
    ————. Title 29—Labor, Chapter XVII—Occupational Safety and Health Administration.
    "Occupational Safety and Health Act of 1970," Public Law 91–596, December 29, 1970.
Note. Government publications are available from Superintendent of Documents, U.S. Government Printing Office, Washington, D.C. 20402.

INDEX

Abrasive blasting, 185–186
Abrasive wheels; see Wheels, abrasive
    grinding
Absorption, skin, 108
Accidents
    analysis; see Safety performance
        measuring
    causes, 11–16, 24–25, 27
    cost of, 1–5, 30–33, 75–76
    elements of, 22–25
    investigation of, 5, 10, 23–27
    measuring; see Safety performance,
        measuring
    off-the-job, 74–78
    prevention of, 6, 8–10, 328
    "proneness," 50–51
    reasons people have, 82
    report forms, 28–29, 32, 34–35
    tool, 330–331
    see also "Near misses" and Injuries
Acid carboys, handling, 273, 275
Acne, occupational, 145–146
Aerosols, 110
Air, working in compressed, 135–137
Air power tools, 362–365
Airborne materials, 108–111, 115–120,
    181
Aisles
    housekeeping, 204, 207, 271
    inspecting, 216
    marking, 222
Alarms, fire, 387–388
Alcohol, use of, 51–52
    countermeasures, 52–53
    interaction with drugs, 51–52

Alcoholism, 51–53
Anthrax, 141
Attitudes, safety, 38, 57
Automation, 322–323
Awards, safety, 68
    "clubs," 64–65, 200
Axes, 344–345

Bags, piling, 230, 271
Battery charging, 239–240
Belts, safety, 188–192
Bends (decompression), 135
Bins, storage, 280
Biological stresses, 140–142
Biomechanics, 138
Biotechnology, 138–139
Bonding, electrical, 274, 277, 384
Bronchial tree, 116
Brucellosis, 142
Byssinosis, 141

Carrying; see Lifting
Chains, 264–269
    storing, 269
Changes, preparing worker for, 46–48
Chemical hazards, 107–108, 386–387,
    396
Chisels, use of, 339–340, 343–344
Clothing, safe work, 194–199, 280
    men's, 198–199
    women's, 196–197
"Clubs," safety recognition, 64–65, 200
"Cold soaking" of metals, 335–336

401

Index

Color, 221–223
  coding, 221–222
Committees, safety, 17, 58–59
Contests, safety, 64, 65–69
Conveyors, 247–248
Cranes, 248–255
  hand signals, 248–250
  inspection, 251–253
Creams, protective, 178
Crowbars, 233
Cutters, metal, 343, 351
Cutting mechanisms, guarding, 301–306
Cylinders, compressed gas, 278–279

Danger zone, definition, 286
Decibel, definition, 174
Dermatitis, 144–146
Dock plates, 245–247
Double insulation, 356–357
Drills, electric, 359
  pneumatic, 365
Driving
  industrial truck, 241–243
    loading, 244–245
  motorized equipment, 257–258
Drugs, use of
  abuse, 52
  countermeasures, 54–56
  interaction with alcohol, 51–52
  policy, 54, 55
Drums
  handling, 230–233
  storing, 274–275
Dusts
  airborne, 108–109, 115–119, 181
  explosive, 279–280, 383–386

Ear protection, 173–176
  damage risk criteria, 121–122
  overcoming objections, 175–176
Efficiency, effect of accidents on, 4–5
Electric equipment, 377–378
Electric power cords, 354–355, 359
Electric power tools, 355–362
Electricity, static, 383–386
Emissions, power truck, 243–244
Employee facilities, inspecting, 217–219

Ergonomic stresses, 138–140
  matching machine to operator, 297–298
Erysipelas, 142
Evacuation, emergency, 397–398
Explosive-actuated fastening tools, 365–368
Eye care
  laser damage, 130
  protective equipment, 153–165, 226

Face protection, 158–159, 161–164
Fatigue, metal, 335–336
Feeding methods, machine, 313–316
Files, use of, 341–342
Finger protection, 176–178, 225–226
Fire brigades, 394–395
Fire causes, 371, 379
  electrical, 377–378, 383–386
  friction, 378
  heaters, 381–382
  ignition, spontaneous, 383
  smoking and open flame, 381, 382
  welding, 380–381
Fire chemistry, 371–373
Fire extinguishers, 388–393
  job analysis, 93
Fire prevention, 398
  during construction, 396
  inspection, 373–376, 393
    checklist, 375
  working with employees, 376, 388, 393
Fire protection, 369–370
  color coding, 221, 223, 393
  plant, 212
Fire pyramid, 371
Fires, classes of, 388–389
First aid
  courses, 71
  reports, 33–37
  statistics, 37
Fittings, wire rope, 264–265
Flexible shaft tools, 365
Floors, inspecting, 215–216
  loading, 245
Foot controls, safeguarding, 315–316
Foot protection, 165–169, 226
  overcoming objections, 168–169
Forming or bending mechanisms, guarding, 308–313

Four-point method of JIT, 96–100
Fueling, industrial truck, 240, 381
Fumes, 109, 119–120, 362
Fungus infections, 140–141

Gas masks, 182–183
Gases, 110, 383
Gasoline power tools, 365
Goggles, 154–161
    cleaning, 160–161
Grounding, electrical, 24, 274, 384–386
    tools, 357–359
Group feeling, 43–44, 56
Guards and safeguards
    acceptable, 287–289
    automatic devices, 319–322
    benefits of, 285
    built-in, 282, 294–296
    construction materials, 288
    definitions, 286
    design, 287–296
    fixed, 317–318
    interlocking, 318–319
    location of, 289
    lockouts, 325–327
    machine, 316–322
    maintenance, 324–328
    mechanisms, 297–316
    openings, size of, 288, 289, 293
    point of operation, 286, 289, 290–
        292
    power press, 308–316
    power-transmission apparatus, 287,
        288, 298–301
    principles, 283–285
    training, 323–324
    types, 290–294

Hair, long, 196–199
Hammers, use of, 352–353
Hand protection, 176–178, 225–226
Handles, tool, 233, 338
Hatchets, 345–346
Hats, safety, 169–173
    display of, 62
Hazards
    controlling, 13, 57, 151–153
    health; see Industrial hygiene
    identifying, 12–16, 92

Hazards—Continued
    machinery, 298
    marking, 221–223
    see also Inspections
Head protection, 169–173
    overcoming objections, 172–173
Hearing protection, 173–176
    damage risk criteria, 121–122
Heat controls, 135
Heat cramps, 134
Heat exhaustion, 134
Heatstroke, 133
History of safety movement, 7–8
Home study courses, 104
Hoods, air supplied, 163
Hooks
    crane, 248–249
    hand, 234
Hose, air, 362–363
Housekeeping, 205–209
    construction, 204–205
    fire-safe, 386–387
    industrial, 201–220, 279
    sources of help, 209–212
    see also Inspections
Humidification, 386
Hygienist, industrial, 106

Ice picks, 348–349
Impact tools, pneumatic, 363–364
Industrial hygiene, 106–150
    airborne materials, 108–111
    biological stresses, 140–142
    chemicals, 107–108
    controls, 148–150
    definition, 106
    dermatitis, 144–146
    ergonomics, 138–140
    operating procedures, 146–150
    particulates, 115–121
    physical agents, 121–128
    pressures, atmospheric, 135–137
    solvents, 111–115
    temperature extremes, 130–135
    threshold limit values, 142–144
    visible radiation, 128–130
Information sources, 20–22
Ingestion of harmful compounds, 108
Inhalation of contaminants, 107

Index

Injuries
  incidence rate, 37
  investigating, 25–26
  measuring, 33–37
    calculating rates, 67, 77
  off-the-job, 37, 74–78
  sources, 20–22, 224, 281
  statistics, 15–16, 75
Inspections, 16–19, 373–376
  cranes, 251–253
  housekeeping, 212–220, 393
  rating sheets, 17, 213–214, 375
  tools, 334
Instruction
  classroom, 101–103
  giving, 83–86
  home study, 104
  job (JIT), 96–101
  on-the-job, 86–89
  programmed, 104
  see also Training
Insurance costs, 3, 30–33
Interlock, definition, 286

Job instruction training (JIT), 96–101
Job safety analysis (JSA), 13, 88–96

Knives, hand, 346–348
  injuries, 16

Lasers, 129–130
Leadership qualities, 44–46
Leg protection, 169
Legal requirements for safety, 19
  see also Occupational Safety and
    Health Act
Lesson plans, 102–103
Life lines, 190–192
Lifting, proper method of, 227–233
Liquids, handling hazardous, 273–278,
  381, 383–386
Lockouts, machine, 325–327
Low voltage tools, 356
Lungs (cutaway drawing), 116

Machines
  guarding; *see* Guards and safeguards
  inspecting, 219–220
  matching to operator, 297–298

Materials handling, 224–280, 297, 313–
  315
  combustible solids, 279–280
  equipment for, 233–258
  gas cylinders, 278–279
  liquids, hazardous, 273–278
  manual methods, 225–233
  ropes, chains, slings, 248, 252–253,
    258–269
  storage, 269–273
Mechanical power-transmission apparatus, safeguarding, 298–301
Medical
  costs, 3
  reports, 33–37
Meetings, safety
  departmental, 71–72
  how to hold, 73–74
  production huddle, 72–73
  tailboard conference, 73
Metatarsal guards, 168
Microwave radiation, 125
Mists, 110, 181
Morale, effect of accidents on, 6
Motivation
  at meetings, 73
  background, 38–39, 57
  principles, 39–44, 56
Motorized equipment, 257–258
  see also Trucks

National Safety Council, 7–8
  Congress, 153
  posters, 60–61
  publications, 21–22, 71, 79, 287, 400
"Near misses," 2–4, 14, 25
Nailers and staplers, air-operated, 364–
  365
New worker, starting, 85–86
Nip-points, safeguarding, 306–308
Noise, 121–122, 298
  see also Hearing protection

Occupational Safety and Health Act,
  8, 19, 33, 150
  fire extinguisher mounting, 393
  guarding and safeguarding, 287, 289,
    294, 311
  housekeeping requirements, 203–205

Occupational Safety and Health Act—
Continued
reports, 28–29, 34–35
tools and equipment, 329
Off-beat safety ideas, 62–64
Off-the-job safety, 74–78
integration with job, 79–81
On-the-job training, 86–89
Over-the-shoulder coaching, 87–89
Oxygen-deficient atmospheres, 137, 280

Parts per million, 144
Personal protective equipment, 151–200
color coding, 221–222
paying for, 200
recognition "clubs," 64–65
Pinch point, definition, 286
safeguarding, 306–308
Planes, wood, 346
Pliers, use of, 351
Pipelines, 277
color code for, 222–223
Pneumoconiosis, 181
Point of operation
definition, 286
protection, 289, 314
Policies, drug, 54, 55
Posters, safety, 60–61
Power presses
definitions, 286
guarding, 308–311
locations, 289
primary and secondary operations,
312–316
Preventive maintenance, 18
Problems, workers with, 48–51
Property damage costs, 3
Public relations, accident effect on, 6
Pumps, maintaining, 148

"Q fever," 141–142

Radiation
human body heat, 131–132
ionizing, 123–125, 192–194, 396–397
nonionizing, 125–128
visible, 127, 128–130
Railroad cars, 255–256, 277–278
Reminders, safety, 61–62

Respiratory protective equipment, 179–
187, 226
cleaning, 187
overcoming objections, 187
Respiratory tract infections, 142
Rollers, moving, 233
Ropes
fiber, 190–192, 258–262
hitch efficiency, 191
hoist, 248, 252–253
wire, 263–264
Rotating mechanisms, safeguarding,
298–301

Safeguards, definition, 286
see Guards and safeguards
Safety movement, growth of, 7–8
Safety performance
measuring, 20–22, 27–30, 37, 67, 77–
78
reporting, 33–36
Safety programs, 2, 8–10
purpose, 1
Salt intake, 134
Sanders, belt or disk, 361
Saws
electric, 359
hack, 341
hand, 344
safeguarding, 301–305
Scaffolds, 23–24
Screw drivers, 351–352
Screw or worm mechanisms, safeguard-
ing, 308
Securely fastened, definition, 286
Security, job, 42–43, 47
Shoes, safety, 165–169
Shovels, 234–235
Skin diseases, 144–146
Slings
chain, 266–267
rope, 262–263
Smokes, 109
Soldering irons, use of, 361–362
Solvent hazards, 111–115, 377–378
"safety solvents," 114–115
Spectacles, protective, 155–159
Sprinklers, fire, 223, 396
height above storage, 270, 387

Storage facilities, 269–280
  fire hazards, 378–380
  inspecting, 216–217
Stresses, environmental, 106
Structures, inspecting, 214–215
Suggestions, employee, 48, 69–71
Sunstroke; *see* Heatstroke
Sweating, 131

Talking to reduce emotions, 50
Tank cars, unloading, 277–278
Tanks, entering, 137, 146–147, 183–184, 280
Taps and dies, use of, 340
Television, closed circuit, 104–105
Temperature extremes, 130–135
  effect on body, 132–135
Tenosynovitis, 140
Threshold limit values, 142–144
Tin snips, use of, 342–343
Tongs, use of, 351
Tool
  belts and bags, 332–333
  boxes, 330
  dressing, 336–338
  handles, 233, 338
  tempering, 334
Tools, hand and portable power
  controls, 330–333, 334
  electric, 355–362
  maintenance, 329–330, 334–338
  materials handling, 233–235
  personal, 329, 332
  safe-ending, 336
  safe practices, 329–330
  selection and storage, 329–330, 331–332, 355
  spark-resistant, 280, 353–354
  use of, 332–333
    hand, 338–354
    power, 354–368
Torso protection, 178–179

Training, 9–10, 41–43, 82–105, 323–324, 330–331
  limitations, 285
Trucks
  four-wheeled hand, 236
  highway, 257
  motorized, 257–258
  powered hand, 236–238
  powered industrial, 238–245
  two-wheeled hand, 235–236
Tuberculosis, 140

Unsafe acts and conditions; *see* Accidents, causes

Vapors, 110
Valves, maintaining, 148
Ventilation
  combustion byproducts, 243–244
  fumes, 119–120
Vibration, mechanical, 139–140
Violations of rules, 14
Visual aids, 71–72, 104–105
Visible radiation, 127, 128–130

Welding
  filter lens shade numbers, 157
  helmets, 162–163
  see also Fumes
"White fingers," 139
Work, value of, 40–42, 56
  group feeling, 43–44
Wheels
  abrasive grinding, 12–13, 303–305, 360–361
  air powered, 363
  buffing and polishing, 305
  wire brush, 305–306
Wrenches, use of, 349–350

Yards and grounds, housekeeping, 220